TELEPHONE VOICE TRANSMISSION
Standards and Measurement

WINSTON D. GAYLER, P.E.
Consulting Engineer

PRENTICE HALL
Englewood Cliffs, New Jersey 07632

Library of Congress Cataloging-in-Publication Data

Gayler, Winston.
   Telephone voice transmission: standards and measurement / Winston
D. Gayler.
     p. cm.
   Bibliography: p.
   Includes index.
   ISBN 0-13-902776-9
   1. Telephone—Standards.   I. Title.
TK6169.G39 1989
621.385—dc19

                                               88-32735
                                             CIP

Editorial/production supervision
  and interior design: Cristina Ferrari
Cover design: Wanda Lubelska Design
Manufacturing buyer: Mary Ann Gloriande and Robert Anderson

The publisher offers discounts on this book when ordered
in bulk quantities. For more information, write:

     Special Sales/College Marketing
     Prentice-Hall, Inc.
     College Technical and Reference Division
     Englewood Cliffs, NJ 07632

Printed in the United States of America

10 9 8 7 6 5 4 3 2 1

ISBN 0-13-902776-9

Prentice-Hall International (UK), Limited, *London*
Prentice-Hall of Australia Pty. Limited, *Sydney*
Prentice-Hall Canada Inc., *Toronto*
Prentice-Hall Hispanoamericana, S.A., *Mexico*
Prentice-Hall of India Private Limited, *New Delhi*
Prentice-Hall of Japan, Inc., *Tokyo*
Simon & Schuster Asia Pte. Ltd., *Singapore*
Editora Prentice-Hall do Brasil, Ltda., *Rio de Janeiro*

To the employees, customers, and suppliers
of the former Vidar Corporation of Mountain View, California.

CONTENTS

CHAPTER 2   LEVELS, LOSS, AND FREQUENCY RESPONSE   33

CHAPTER 6   LONGITUDINAL BALANCE   164

PREFACE

This book is concerned with the analog transmission of voice-frequency telephone signals. The topics covered are those of interest to designers of electronic telephone equipment. Such equipment includes individual telephone sets, key telephone systems, private branch exchanges, local and long-distance switching offices, and carrier system channel banks.

In the past, telephone technology was the sole province of the former Bell System and a few other firms. Their engineers received in-house training and there was little need for outside dissemination of technical literature. In recent years, the Bell System divestiture and the advent of new common carriers and customer-provided equipment have restructured the telephone industry. New markets have been created and new firms started to supply these markets. The technical personnel in these firms now need literature in the field of telephony. The Electronics Industries Association, the Institute of Electrical and Electronics Engineers, and others have responded by publishing standards for the telephone industry. These standards, however, contain little tutorial matter and are often difficult for the novice to interpret.

Technical writers have also responded. However, with the evolution of the telephone network toward digital encoding, transmission, and switching of voice communications, there is a tendency to overlook the analog technology still used in much of the network. This book covers analog voice-frequency techniques and can be read as a simple-language supplement to the formal documents of the EIA, IEEE, FCC, and others.

Although analog parameters are emphasized, this is not done at the neglect of digitized voice. When voice signals are digitized, transmission impairments such as quantizing noise and aliasing arise. These digital effects and their measurement are also covered.

This book is directed primarily toward practicing design engineers and their technician assistants who need to learn basic telephone voice transmission. Developed for over a century and almost entirely within one company, U.S. telephony has a vocabulary of its own. The demystifying of this vocabulary is often all that is needed before experienced engineers can contribute to telephony design projects. They recognize the underlying engineering principles and are able to apply circuit analysis and synthesis techniques learned in other fields. This book will also be of value to engineers and technicians who are users or specifiers of telephone equipment or telephone test instruments. Engineering students may wish to read this book as a supplement to a formal text in communications theory.

To use this book effectively, you should have a basic background in electronics. This includes basic circuit analysis skills using operational amplifiers, comparators, transformers, and discrete transistors. The mathematics used is limited to basic algebra and trigonometry. Laboratory technicians and junior-year electrical engineering students should be able to follow the text fully.

Much of the material presented here was gathered during my work as a designer of telephone switching and transmission equipment. I have included tips that may help you avoid some of the pitfalls of telephone equipment design.

Chapter 1 provides a brief history and overview of the telephone network. This material can serve as an introduction for the novice or as a review for those already in the industry. The chapter also contains an introduction to signaling, which is an important companion subject to transmission.

Chapters 2 through 7 cover the transmission parameters that affect voice-frequency performance of telephone circuits. The measurement method is given for each parameter and the applicable measuring instruments described. Throughout these chapters you will find measurement and design examples that illustrate the theory. Finally, the industry standards that specify performance are summarized for each parameter. When available from industry surveys, the network's actual performance is compared with the published objectives.

The main topics by chapter are:

Chapter 2:   Signal levels (voltage, power, dB, dBm), impedance, transmission level points, loss plans, trunk design, level stability and tracking, frequency response.

Chapter 3:   Noise (steady and impulse), crosstalk.

Chapter 4:   Hybrids, two- and four-wire circuits, transhybrid loss.

Chapter 5:   Echo return loss, hybrid balance networks, build-out components, office balance.

Chapter 6:   Longitudinal balance.

Chapter 7: Distortion (intermodulation, quantizing, aliasing), jitter, hits, and dropout.

Chapter 8 discusses telephone industry standards and regulations. Compliance with FCC rules is a legal requirement for the sale in the United States of customer premises equipment. Voluntary standards for telephone equipment are published by the Electronics Industries Association. Designs that meet EIA standards can usually be assured of acceptable performance and of compatibility with other equipment. Other standards are issued by the Institute of Electrical and Electronics Engineers, Bell Communications Research, and the Rural Electrification Administration. Chapter 8 deals with the applicability of these standards and tells how to obtain copies.

ACKNOWLEDGMENTS

I wish to thank James Alinsky for his review, from the reader's perspective, of the complete manuscript. I am also grateful for the engineering reviews made by Gene Acton, Bert Dotter, Steve Grimes, Ron Hagen, Brad Helliwell, and Jim Hon. Thanks also go to Heidi Tressler for assistance with correspondence and manuscript preparation.

A most grateful acknowledgement is due Suzanne Moore for dedication and attention paid to detail in typing the full manuscript.

Cupertino, California   Winston D. Gayler

REFERENCE TABLES

1

OVERVIEW of the TELEPHONE NETWORK

For the purposes of this overview, we will divide the telephone network into three parts: transmission, switching, and signaling. *Transmission* involves the transport of voice and data signals between points on the network. The technology used can range in complexity from that of a pair of wires to that of a multichannel satellite link. The signals are transmitted at their original voice frequencies, or the signals are used to modulate carriers at higher frequencies. Voice signals can also be digitally encoded and transmitted as bit streams on wire pairs, microwave links, and optical fiber. In whatever form they take, the signals are subjected to transmission impairments such as noise and distortion. The measurement of those transmission impairments that occur at voice frequencies is the subject of this book.

The other two parts of the telephone network, switching and signaling, are covered in this chapter's overview. *Switching* has to do with routing telephone calls from place to place in the network. Telephone switching offices can be thought of as the nodes of the network, while transmission facilities are the links. At a telephone *switch,* transmission links known as *trunks* are connected together to form part of a *built-up connection* through the network.

Signaling refers to call setup information that is sent from switch to switch. For example, a telephone number that you are calling will have to be sent to the appropriate switch to tell it which phone to ring. Signaling can be sent on the trunk used for the associated conversation, or it can be sent on a separate link.

1-1 AN INTRODUCTION TO TELEPHONY

A Local Example

The example of a local telephone call will illustrate the combination of transmission, switching, and signaling. It will also introduce and define some of the telephone terminology used throughout the book.

The local network.   A local telephone network (Fig. 1-1) comprises a *central office* and a number of subscriber (customer) telephone sets. The telephones are connected to the central office via *subscriber loops*, also known as *local loops,* simply *loops,* or more recently as *end-user access lines.* A subscriber loop is usually a twisted pair of copper wires; however, more sophisticated implementations such as subscriber carrier are common. *Subscriber carrier* systems multiplex the signals for a number of subscribers on one pair, thus providing *pair gain.*

The central office is a telephone switch and may also be called a *local switch, local office,* or *end office.* (The term *class 5 office* has been replaced by *end office.*) We will use the terms *Central Office* (CO) and *End Office* (EO) interchangeably in this book to designate a switch to which subscriber telephones attach. The CO connects subscriber loops together to complete a local call. Calls to distant telephones are switched out of the office via interoffice trunks.

Returning to the local call, we now examine it in detail (Fig. 1-2). The call *originates* at telephone set A, switches through the local CO, and *terminates* at telephone B. The CO in this example is a digital switch.

Seizing the CO.   Transformer T_1 in line circuit A provides a −48-V *battery feed* to telephone A via the associated loop. The two loop conductors are designated *Ring* and *Tip* (R&T). Normal polarity is negative on ring. The transformer windings and the loop conductors are both ac- and dc-balanced with respect to ground.

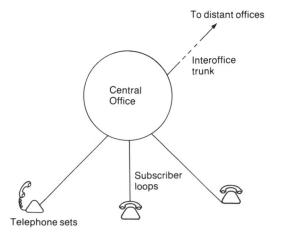

FIGURE 1-1   Local telephone network.

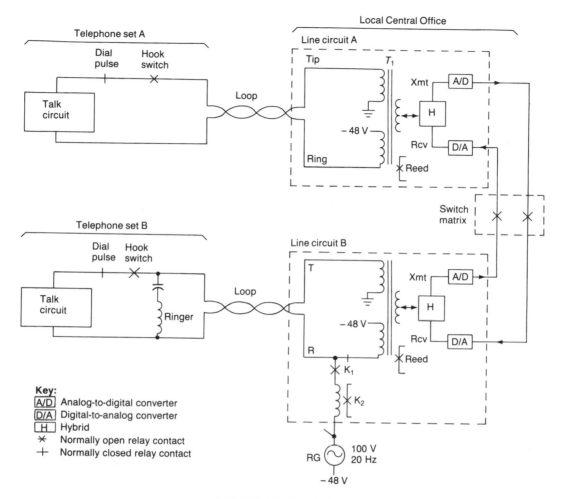

FIGURE 1-2 Local telephone call.

When the subscriber lifts the handset to make a call, the *hook switch* contacts close, completing a dc path from -48 V through T_1, the loop, the telephone, and back to ground. The *loop current* that flows powers the telephone's talk circuit. The loop current is also detected in line circuit A, thus *seizing* the CO. We have shown a reed relay contact mounted near T_1's core as the means of current detection.

As an example of signaling, note that lifting the handset changes the *signaling state* that is sent toward the CO from *on-hook* (idle) to *off-hook* (busy).

When the hook switch closes, a Voice-Frequency (VF) transmission path is established between the telephone's talk circuit and the CO's line circuit. (Since this path carries both directions of transmission, it is known as a *two-wire* circuit.)

The path couples via T_1's secondary to a *hybrid circuit* which separates the two directions of transmission into *Transmit* (Xmt) and *Receive* (Rcv). (A circuit with separate transmit and receive subcircuits is called a *four-wire* circuit.) The transmit side is digitized in an Analog-to-Digital (A/D) converter, then it is presented to a four-wire switching matrix. Digitized signals coming from the matrix are converted to analog via a Digital-to-Analog (D/A) converter.

The central office is run by software which is executing in a computer. This scheme is known as *stored program common control*. When the CO is seized by line A, the stored program causes a dial tone to be connected to the line circuit via the matrix. (In a digital office, tones such as the dial tone are simply stored in a memory as a bit pattern.) At the same time, the program assigns a dial pulse *register* (receiver) to monitor the line and accumulate the number dialed from telephone A.

Dialing.   Each pull of the telephone's dial generates 1 to 10 dial pulses, corresponding to the digits 1 to 0. A pulse occurs when the dial pulse contacts open momentarily. Dial pulses are nominally sent at 10 per second and at *60% break*. (That is, the break time equals 60% of the make-plus-break time.) The break time is made longer than the make time to compensate for loop capacitance, resulting in improved performance of the line circuit's dial pulse detector. The *interdigit* time (end of last break of one digit to start of first break of next digit) should be at least 700 ms. Dial pulses are another example of *signaling*.

The common control gathers the dialed digits from the dial pulse register and *translates* the assembled *directory number* into a called *equipment location*. The called equipment location in this example is line circuit B. If line B is busy (its reed contacts found closed), the common control causes a busy tone to be returned to line A in a manner similar to sending a dial tone.

Ringing.   If line B is idle, the common control returns an *audible ringing* tone to line A. The common control also rings telephone B by causing relay K_1 in line circuit B to operate. This transfers the ring lead from -48 V to a common Ringing Generator (RG) external to the line circuit. (*Common* refers to the fact that the RG is shared by a group of line circuits.) The ringing generator provides approximately 100 V ac at 20 Hz superimposed on -48 V dc.

In the telephone, there is a *ringer* that is ac-coupled across R and T. The 20-Hz ringing current passes through relay K_2's coil, relay K_1's contacts, the ring lead, the ringer and its capacitor, the tip lead, ground, and back to the ringing generator via the -48-V battery. Note that only ac flows through this path. Ringing is *interrupted* by operating and releasing relay K_1 with the desired timing, usually 2 seconds on and 4 seconds off.

Ring trip.   Ring trip relay K_2 is a special relay (or electronic circuit) that operates on dc but not ac. It does not operate on ringing current. When the called party answers telephone set B, its hook switch closes, and a dc load is provided to the loop. Since the ringing voltage is superimposed on -48 V, dc flows and K_2

operates. This alerts the common control that B has answered. If the called party answers during the *silent interval* (i.e., while K_1 is released), the reed contact closes to alert the common control. Upon detecting an answer at telephone B, the common control stops ringing and establishes a path between line circuits A and B as shown in Fig. 1-2. The two parties can finally converse.

Ringing circuitry also exists on line A; it has been omitted from Fig. 1-2 for simplicity. The circuitry shown in the figure is representative of actual designs, but it is certainly not the only way to design a central office. Whatever telephone set and central office designs are used, the interface at the tip and ring leads must meet industry standards. This standardization allows telephones from a variety of manufacturers to work over a wide range of loop conditions and with central offices that also come from several manufacturers.

The Telephone Network

The telephone network (called simply *the network* by workers in the field) is over 100 years old. During most of that time, *the network* and *the Bell System/AT&T* were synonymous. Bell had a monopoly not only on telephone service, but also on the technical terminology used by telephone engineers. One of this book's objectives is to decode the industry jargon relating to voice-frequency transmission.

The network has been changing lately and is no longer the monolith it once was. These changes are discussed in the next section. In telephony, much of the new is based on much of the old. To understand today's network, you should know something of its past.

1-2 EVOLUTION OF THE NETWORK

1-2-1 Steady Progress

From its beginning and through the mid-1960s, the telephone network experienced slow but steady technological progress. By that time, dial phones and direct distance dialing had become the norm. Digital carrier (multiplex) equipment was introduced into the transmission plant in 1962 and the first electronic (stored program control) central office was put in service in 1965. Touchtone service was offered in 1963 and the Trimline* telephone was introduced in 1965. Modems (known as Dataphone* equipment then) came into use to connect computers over phone lines. Still, as far as most subscribers were concerned, the big news of the decade was color phones and long cords!

The Network

The telephone network circa 1965 is pictured in Fig. 1-3. Then, as now, individual subscriber loops connected to the central office. *Local calls* were

---

* Trimline and Dataphone are registered trademarks of AT&T.

switched within COs as described earlier in this chapter. *Toll* (long-distance) *calls* were switched out of the CO and routed to a *toll office* via a Toll-Connecting Trunk (TCT). Calls that routed between toll office did so over Intertoll Trunks (ITTs).

Note the existence of a switching hierarchy with local COs designated *class 5*, and toll offices designated *class 4, class 3,* and higher. With *hierarchical routing,* calls route from the local CO to the next-higher-class office; then, if necessary, to even higher-class offices before starting back down the hierarchy toward the far-end CO. Typically, however, calls skip the higher levels of the hierarchy and route directly between lower-level offices.

By examining Fig. 1-3, you can observe that a toll connection comprised two loops, two TCTs, and possibly several ITTs. This gave rise to different grades of performance for transmission components. Since impairments can be additive when analog circuits are connected in tandem, and since several ITTs may have

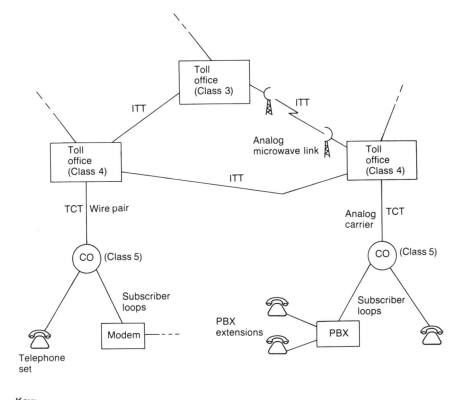

Key:
CO   Central Office
ITT   Inter-Toll Trunk
PBX   Private Branch Exchange
TCT   Toll-Connecting Trunk

FIGURE 1-3 The network circa 1965.

been connected in tandem for a call, ITTs were designed with the highest-performance equipment—*toll grade*. Toll circuits are shared by many users, so the cost of this high performance was distributed over many calls.

At the other performance extreme were subscriber loops. Any performance enhancements made to all loops on a network-wide basis were very expensive due to the large quantity of loops involved. However, any call used only two loops, so loop transmission performance did not need to be as high as that of toll equipment.

The performance required of toll-connecting trunks fell between that of ITTs and loops. Equipment designed for loop and TCT applications was designated *exchange grade*.

Facilities

The transmission facilities of 1965 included wire pairs (for loops and TCTs) and analog wire-line carrier and analog microwave links (for TCTs and ITTs). Digital carrier had just been introduced for use in the exchange (nontoll) plant. *Open wire* (steel or copper-clad steel conductors tied to glass insulators) was still in use at the same time that the Telstar satellite was demonstrating international satellite communications.

Loaded and nonloaded cable. When wire pairs were bound in cables for use at voice frequencies, the pairs were often loaded. *Loading* is the insertion of series inductance for the purpose of reducing attenuation. Practical loading schemes use *lumped* inductance inserted into the pair at equal intervals. The combination of distributed shunt capacitance and lumped series inductance turns the pair into a low-pass filter. The cable capacitance and the loading inductance and spacing are controlled so that the cutoff frequency of *loaded cable* is above the voice band. One of the most common loading schemes uses 88-mH inductors placed every 6000 ft. When used on cable with a standardized capacitance of 0.083 μF/mi, H88 loading provides a cutoff frequency of about 3.5 kHz.

In loaded cables, the lumped inductance is provided by a *load coil*. The load coil's inductance (and its slight series resistance) is split evenly between the tip and ring conductors so as to maintain the cable's longitudinal balance.

In contrast the loaded cable, *nonloaded cable* has greater attenuation in the voice band and a sloped frequency response—the high frequencies are attenuated relative to the low frequencies. These characteristics limit the use of nonloaded cable to short runs, typically under 18,000 ft. Note, however, that most subscriber loops are short enough to use nonloaded pairs.

Switching

The switching offices in use in 1965 were a combination of *Step-by-Step* (SXS), *Crossbar* (Xbar), and the new electronic 1ESS* switching office. All were *electromechanical* switches, meaning that the actual switching element resembled an electromechanical relay with its electric magnet operating a set of mechanical

* ESS is a trademark of AT&T.

contacts. The 1ESS switching office uses a sealed reed contact; the others use larger unsealed contacts.

A SXS office is under *direct control,* meaning that each dial pulse from the subscriber operates the switching elements directly to set up a path. The Xbar and 1ESS switching offices use a *common control* that is shared by all calls. The common control interprets dialed digits before operating the switch matrix. The Xbar's control is implemented in relay logic, while the 1ESS switching office uses a stored program (software) for control.

Ownership

One of the largest differences between the network of 1965 and the one of today is ownership. The Bell System owned *everything* in Fig. 1-3—even the modem. Subscribers were forbidden from connecting their own phones or any other device to the network.

Independent telephone companies.   In actuality, ownership of the 1965 network was shared by AT&T and the *independent* telephone companies. An independent operates like a Bell company in that it too has a monopoly on local service in its geographical area. In 1965 the independents also operated small toll networks within their service areas, but connected with the Bell system for handling longer-distance calls. General Telephone, United Telephone, and Continental Telephone are examples of large independents. Independent telephone companies should not be confused with interconnect companies, which are a relatively recent development. Since all companies are independent following divestiture, "independent" now means non-Bell.

1-2-2 Customer-Provided Equipment

The Carterphone Decision

The *Carterphone decision* of 1968 is usually considered the beginning of rapid changes in the network. Manufactured by the Carter Electronics Corporation, the Carterphone was a device that connected a two-way radio system to the network via a subscriber's loop. This was forbidden by the phone company. However, the Federal Communications Commission (FCC) ruled that connections to the Carterphone and other Customer-Provided Equipment (CPE) were to be allowed. Following this ruling, *interconnect* companies were formed to sell, lease, and install CPE.[19] ("Interconnect company" is another term that has lost much of its original meaning since divestiture.)

Protecting the Network from Harm

Protective devices.   When the FCC ruled in the Carterphone case, it allowed the telephone companies to install a protective device between the loop and

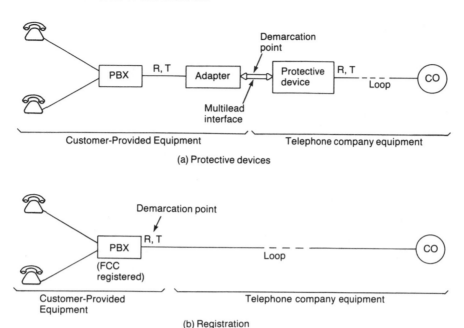

FIGURE 1-4   Network protection from customer-provided equipment.

the CPE (Fig. 1-4a). Bell claimed that these devices would protect the network from harm that might come from the CPE, such as overvoltages and mutilated dial pulses. Bell charged a monthly fee for the protective devices and designed them with a nonstandard, multilead interface at the Bell/subscriber demarcation point. This forced the subscriber to lease from its interconnect company an adapter to convert the multiple leads back to the standard ring and tip.

Equipment registration.   In the early 1970s the FCC ruled on a more sensible approach to the question of network harm by CPE. Protective devices were no longer required, but CPE had to meet certain technical requirements and be registered with the FCC (Fig. 1-4b). This led to the development of Part 68 of the FCC Rules and Regulations.[1,2,3] Part 68 contains the technical requirements for registration of *terminal equipment*. These requirements deal with equipment conditions that could harm phone company personnel (i.e., excessive voltage), that could interfere with *another* user's calls, or that could cause incorrect billing. Except for these limited considerations, Part 68 does not address equipment performance or even equipment operability. These items are left to the marketplace to judge. As Fig. 1-4b shows, equipment registration allows the customer/telephone company demarcation point to fall very naturally at the tip and ring leads as they enter the customer's premises.

1-2-3 Digital Switching

Recall that digital trunk carrier was introduced in the 1960s. This system uses Pulse Code Modulation (PCM) and multiplexes 24 voice channels and their associated signaling on two wire pairs (one for each direction of transmission). Repeaters (actually pulse regenerators) are placed at approximately 1-mile intervals. The line rate is 1.544 megabits per second. At each end of the digital line is a terminal, called a *channel bank,* that contains A/D and D/A converters and other circuitry for each voice channel. The early channel banks are designated D1, and the whole system is called *T1 carrier.*

The D1 terminals were replaced by an improved, toll-trade terminal designated D2. The D2's voice-frequency characteristics are the standard still in use today. Later channel bank implementations using the D2 format are labeled in the sequence D3, D4, ... *Digital multiplexers* are available that multiplex subgroups of 24 channels onto digital lines operating at rates higher than do T1 lines. These higher-rate lines are designated T1C, T2, T3, and up.

The early 1970s saw the introduction of digital carrier on subscriber loops. Typical systems allow a group of 24 or 96 subscribers to be served by just a few cable pairs. Digital carrier was initially a stand-alone system, designed into the network on a link-by-link basis. At each switching office, the digitized voice had to be converted back to analog to pass through the switch matrix (Fig. 1-5a).

In 1976, both Bell and the independent industry introduced digital toll switching offices into the network. Digital local offices followed quickly. These switches use the same bit rates and encoding methods as those of digital carrier. This compatibility allows direct digital connection between switching and trans-

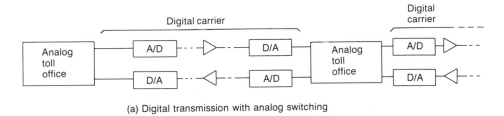

(a) Digital transmission with analog switching

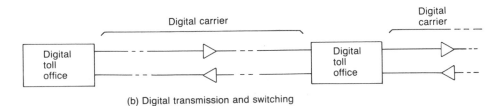

(b) Digital transmission and switching

FIGURE 1-5 Synergy of digital transmission and switching.

mission elements. Direct digital interconnection eliminates back-to-back A/D and D/A conversions when calls route through switching offices. In areas with heavy T-carrier use, digital switches are a natural replacement for old analog offices. Once digital switching is in place, the use of digital carrier is an obvious choice. This synergy is depicted in Fig. 1-5b.

1-2-4 Other Common Carriers

Other Common Carriers (OCCs) are those other than AT&T. In 1970 the first OCC (also called in this case a *specialized common carrier*) was authorized by the FCC to provide high-volume leased private lines for business customers between major cities. The Bell companies were required to interconnect their customers to the OCCs. The other common carriers next offered long-distance service to individual business and residential subscribers.

Line access.   So that subscribers could reach the OCC's networks, Bell first gave the OCCs *line-side access* to Bell's local COs (Fig. 1-6a). In this scheme, the OCC subscriber first dials a seven-digit access number. The access number is a standard Directory Number (DN) appearing in the subscriber's CO or in one nearby. The line equipment associated with the DN connects via a loop to a switch on the OCC network.

The caller next dials a secret access code into the OCC switch. The caller must use a DTMF (touchtone) telephone. The OCC checks this code for validity and records it for billing purposes. Finally, the caller dials the called number (again using DTMF only), and the OCC routes the call to a line appearance on the distant Bell CO.

Note the required use of DTMF—dial pulse signaling will not work. This is because dial pulses are not sent backward over the loop from the local Bell CO to the OCC switch. Tones *will* traverse this path. As a related deficiency, answer supervision does not pass over the loop at the distant Bell CO. (*Answer supervision* is a change in trunk signaling state that tells the calling office that the called party has answered.) Without answer supervision, the OCC has to guess if and when the called party answers and bill accordingly—sometimes in error. Note that under the original OCC *line-side access* scheme, an OCC call involves four subscriber loops, while an AT&T call involves only two. Thus all other things being equal, an AT&T call had better transmission, more accurate billing, and fewer dialed digits than the same call on an OCC. This was because AT&T enjoyed *trunk-side access*. In current terminology, the OCC line access scheme just described is called *Feature Group A*.

Trunk access.   To eliminate the deficiencies of line-side access, local telephone companies are now required to provide trunk access to other common carriers (Fig. 1-6b). This scheme, where AT&T and all OCCs have trunk-side access, is known as *equal access* or *Feature Group D*. The local companies are allowed to implement equal access on a realistic but timely schedule.

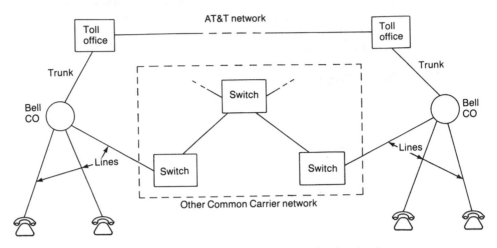

(a) Line access by Other Common Carriers (Predivestiture)

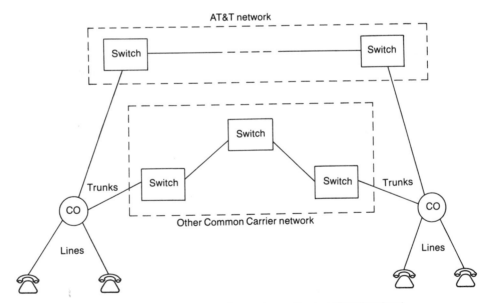

(b) Trunk access by all Interexchange Carriers (Postdivestiture)

FIGURE 1-6 Common carrier access.

The OCCs and AT&T together are called *Interexchange Carriers* (IXCs). With equal access, each customer of the local phone company *presubscribes* to one IXC. When a user makes an IXC call, simply by dialing the called number, the call routes over the presubscribed carrier. If a different IXC is desired, the caller first dials an access number of the form 10XXX, where the XXX selects a particular IXC. The caller then dials the called number. No secret code is needed since the local CO sends the calling number to the IXC office. (Credit card calls still require a secret code.) With equal access, the IXC also receives answer supervision and enjoys better transmission performance. All dialed digits may be sent using dial pulses; DTMF is no longer a requirement.

1-2-5 Bell System Divestiture

On January 1, 1984, AT&T divested itself of the Bell operating companies and part of Bell Telephone Laboratories. The Bell companies were further split into seven *Regional Bell Operating Companies* (RBOCs). Each of the seven RBOCs is a separate corporate entity. The part of Bell Labs that was divested is named Bell Communications Research (Bellcore) and for a minimum period of time is jointly owned by the seven RBOCs.

Postdivestiture, AT&T provides long-distance services and manufactures telephone equipment of all types (from individual telephones to large toll switches). AT&T also sells and leases subscriber equipment but does not provide local telephone service.

Postdivestiture, the Regional Bell Operating Companies provide local telephone service and some short-haul long-distance service. In time, the RBOCs may be permitted to manufacture equipment and provide more long-distance service.

By providing telecommunications research and planning that is of common use to all RBOCs, Bellcore continues one of the functions of the predivestiture Bell Laboratories.

The Postdivestiture Network

Figure 1-7 shows the postdivestiture network circa 1988. There have been major changes since 1965 (Fig. 1-3)—some due to divestiture, others due to technological progress and regulatory changes. There is also new terminology.

Each Regional Bell Operating Company is divided into several *Local Access and Transport Areas* (LATAs). Intra-LATA calls (local or toll) are handled by the RBOC. Inter-LATA calls are handled by interexchange carriers. Within each LATA, there are *end offices* (previously called class 5 offices) and *tandem offices* (previously called toll offices). The letters *TCT* are now an abbreviation for *Tandem-Connecting Trunk* and *ITT* now stands for *Intertandem Trunk*. The IXCs have access to the subscribers in each LATA via *Tandem Inter-LATA Connecting*

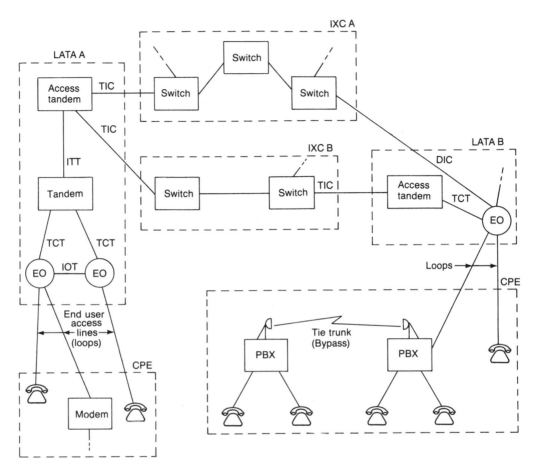

Key:

CPE   Customer Premises Equipment
DIC   Direct Inter-LATA Connecting Trunk
EO   End Office
IOT   Inter-End Office Trunk
ITT   Intertandem Trunk
IXC   Interexchange Carrier
LATA Local Access and Transport Area
PBX   Private Branch Exchange
TCT   Tandem-Connecting Trunk
TIC   Tandem Inter-LATA Connecting Trunk

FIGURE 1-7   The postdivestiture network circa 1988. (Adapted from Bellcore *Notes on the BOC Intra-LATA Networks—1986*.[13])

Trunks (TICs), which connect to *access tandems*. An IXC can also access an end office directly via a *Direct Inter-LATA Connecting Trunk* (DIC).

Subscribers are connected to end offices via *end-user access lines* (subscriber loops). Since customer-provided equipment is no longer the exception, CPE is changed to mean *Customer Premises Equipment*.

Facilities

In the late 1980s, the network makes heavy use of digital transmission on wire pairs, microwave radio, and optical fibers. Still, older analog multiplex is in use. Analog voice-frequency wire pairs are still the norm for subscriber loops; however, *digital and fiber subscriber loops* are in the field-trial stage.

Switching

Digital switching technology is used in switches of all sizes and applications: key telephone systems, Private Branch Exchanges (PBXs), end offices, tandem offices, and IXC offices. A new type of digital switch, called a *Digital Access and Cross-connect System* (DACS), has emerged. The DACS does not switch calls on a call-by-call basis; rather, it is used to reconfigure networks and special service circuits. This is done digitally, on individual circuits or in groups, and with the flexibility of software control. The DACS can be thought of as a smart digital patch bay.

Remote access to switches for maintenance, administration, reconfiguration, and transmission test is common. Routing in the network need no longer be based on the traditional hierarchy of offices. The AT&T network, for example, now uses nonhierarchical routing, which results in more efficient trunk use.

Ownership

In 1988, everyone can own some of the action. In even the simplified network of Fig. 1–7, we have shown two interexchange carriers, one (or maybe two) Regional Bell Operating Companies, plus an assortment of customer premises equipment (owned by the customer or leased from a third party). It is even possible for a customer to own enough of the network to *bypass* the local telephone company completely (see the PBX tie trunk example in Fig. 1-7).

1-2-6 The Integrated Services Digital Network

The telephone network is now evolving toward an *Integrated Services Digital Network* (ISDN). In the ISDN, not only are digital transmission and switching integrated, but services are integrated as well. This means that voice, video, data, and facsimile can all use a standardized digital facility that connects the subscriber to the central office. For the residential and small business subscriber, that digital facility can be the same copper-wire pair now used for analog voice transmission.

The ISDN provides true digital transmission from subscriber to subscriber, eliminating modems that convert digits to VF tones. The ISDN is not a private network or a special service, both of which require special handling by the telephone company. Instead, the ISDN is the evolution of the current worldwide public telephone network. The ISDN is thus an "international end-to-end digital public switched telephone network" and will become a standard service offering.

ISDN Data Rates

Two ISDN data rates are planned for use over copper-wire pairs. The *basic rate* provides two *B channels* of 64 kilobits per second (kb/s) each, plus one *D channel* of 16 kb/s. All channels are bidirectional. The D channel is intended for signaling (call setup information) and low-speed user data. Use of one of the B channels is envisioned for digitized voice, while the second B channel can be used simultaneously for a digital service such as data or facsimile. The three channels have a total capacity of 144 kb/s. The actual line rate will be somewhat higher due to the requirement for framing and other overhead bits. The data are sent bidirectionally on a single subscriber pair. This basic-rate interface between subscriber and CO is known as the *U interface*.

In countries with competitive telecommunications markets, the subscriber and telephone company are free to obtain their equipment from different manufacturers. This requires that the U interface be standardized. Bell Communications Research (Bellcore) has published a proposed U interface standard for use in the United States. The document is Technical Advisory TA-TSY-000393 (see Chapter 8 for information on obtaining Bellcore publications).

The second ISDN data rate is the *primary rate* and is offered for larger users. The primary rate provides 23 B channels of 64-kb/s capacity each plus one D channel, also with a capacity of 64 kb/s. As with the basic rate, the primary rate B channels are intended for user services, while the D channel is used for call setup and control. Notice that the total primary rate capacity (24×64 kb/s) plus an 8-kb/s framing bit channel equals the T1 line rate of 1.544 Megabits per second (Mb/s). This will allow ISDN primary rate interfaces and transmission lines to use much of the hardware already developed for T1 carrier applications.

The ISDN is now moving beyond the field-trial stage. National and international standards organizations are at work defining the ISDN interfaces—and manufacturers are turning out hardware and software designs to implement those interfaces. There is much interest today in the ISDN and there will continue to be so. However, the existing network will evolve into the ISDN over a period of time. During that time, analog and digital portions of the network will coexist. The analog portion will need continued maintenance, and manufacturers will continue to offer products with analog voice-frequency interfaces.

1-3 TRANSMISSION AND SIGNALING

In this section we provide an overview of telephone transmission and signaling, specifically transmission at voice frequencies and the signaling that is associated with individual VF circuits. Furthermore, since the remaining chapters deal with transmission, this overview will emphasize signaling.

1-3-1 Signaling Systems

Circuit-Associated Signaling

Figure 1-8a shows the conventional means of interofffice signaling. Here the signaling associated with each call is sent on the same channel or facility as that of the call's voice transmission. This method is called *circuit-associated signaling*. Since transmission and signaling share the same circuit, one can influence the other. For this reason, we describe a number of circuit-associated signaling schemes. Knowledge of these schemes will be useful when reading the following chapters.

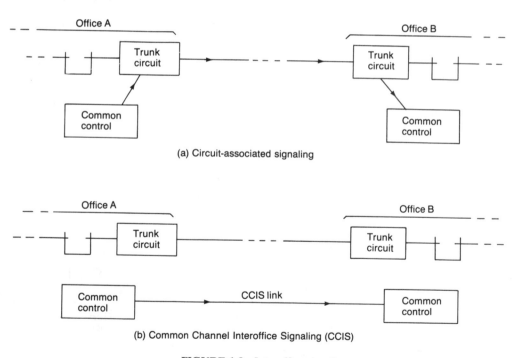

(a) Circuit-associated signaling

(b) Common Channel Interoffice Signaling (CCIS)

FIGURE 1-8   Interoffice signaling.

Common Channel Signaling

Figure 1-8b shows the newer and preferred signaling method, *Common Channel Interoffice Signaling* (CCIS). In this method, a separate high-speed data link is provided for signaling between offices. There are many advantages of CCIS, including faster call setup and the cost saved by eliminating signaling equipment on a per-channel basis. Since CCIS has little effect on transmission, we will conclude our discussion here.

1-3-2 Transmission and Signaling Interfaces

Basic Interfaces

Figure 1-9 shows the various black boxes through which a phone call passes in traversing a switching office. At the heart of a switching office is the switching matrix. The actual matrix elements could be relay contacts, reed contacts, PNPN diodes, field-effect transistors, or logic gates. In general, these elements cannot connect directly to the outside world. Nor is the interface to the matrix standardized.

A *trunk circuit* is a logical and physical block of circuitry that converts the switch manufacturer's proprietary matrix interface to a standardized telephone interface. The matrix, the trunk circuits, and the common control (not shown) make up a *switching office*. (An *office* should not be confused with the building that houses it. Also, one building could contain more than one office.)

A *trunk* is a circuit between offices that carries the voice transmission and associated signaling (if any) for one phone call at a time. A trunk comprises a *facility* plus equipment at each end. Examples of facilities include wire pairs and analog and digital multiplex channels carried on wire pairs, coaxial cable, microwave radio, optical fiber, and satellites. Two or more facility types may be con-

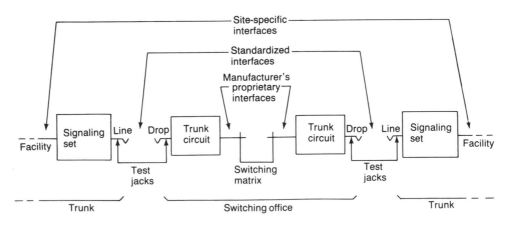

FIGURE 1-9 Basic transmission and signaling interfaces.

nected in tandem to form one trunk. (Note that a *trunk* is not the same as a *trunk circuit*.)

The interface characteristics at the ends of a facility are a function of the facility type, length, wire gauge, and repeater spacing. That is, the interface is site-specific. A *signaling set* is a logical and physical block of circuitry that converts the site-specific facility interface to a standardized telephone interface. (Our broad definition of signaling set includes carrier terminals and channel banks.)

Happily, and obviously part of the plan, the standard interfaces of the trunk circuit and the signaling set match each other. A set of test jacks is often placed at this demarcation point. The jacks' *line* side looks out from the office and toward the trunk to the far office. The *drop* side looks in toward the switch.

1-3-3 Signaling Types

For simplicity in describing the following examples of signaling, we have shown the traditional relay implementations. Alternative electronic designs are common.

Loop Signaling

Loop signaling is the type used between a telephone set and a central office. (Recall the example earlier in this chapter.) Loop signaling can also be used between offices. The basic versions of loop signaling are described next.

Loop-start.   Loop start is the type of loop signaling used by a plain telephone set (Fig. 1-10a). A battery feed of nominally 48 V dc (negative on ring) is supplied by the CO. [We have shown the traditional 400-ohm (Ω) split-winding loop relay.] When the telephone set hook switch operates and closes the loop, the CO is seized—thus *loop-start*. Loop-start signaling is used for subscriber loops and *one-way* PBX trunks.

Ground-start.   Figure 1-10b shows a *ground-start* loop signaling circuit. The CO provides -48 V feed on the ring, but the tip lead is opened by relay K_2 contacts when the circuit is idle. The PBX cannot seize the CO simply by closing the loop. To seize the CO, the PBX must ground the ring lead—thus the description *ground-start*. The ring is grounded via relay K_4 contacts. This operates the CO's battery feed relay (K_1) via the ring winding.

When the CO is ready to receive dial pulses, it grounds the tip (via K_2 contacts and the tip winding of the battery feed relay). Tip-ground is then detected at the PBX by a battery-connected relay coil connected to the tip lead (K_6). The PBX next operates its loop-closure relay (K_3), releases its ring-ground relay (K_4), and operates relay K_5 to disconnect the K_6 coil from the tip lead. Dial pulsing now takes place on a loop basis via K_3 contacts.

For calls in the other direction, the CO seizes the PBX by grounding the tip lead via K_2 contacts. The PBX detects this via K_6 and starts alerting (ringing) the attendant (operator). When the attendant answers, the PBX operates its loop-

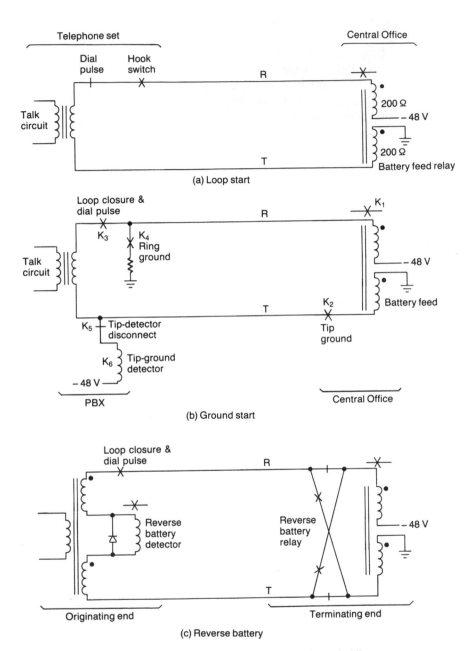

FIGURE 1-10 Loop signaling. All relays shown in idle state.

closure relay (K_3), which causes the CO's battery feed relay (K_1) to operate. The circuit now assumes a loop configuration.

Ground-start signaling is used on two-way PBX trunks, where the elaborate hand shaking just described is used to reduce *glare*. Glare is the simultaneous seizure of both ends of a trunk, usually by parties who did not intend to talk with each other.

Reverse-battery. Figure 1-10c shows a one-way, loop-start trunk that uses reverse-battery signaling. The originating end seizes the terminating end in the manner described previously for loop-start circuits. However, the terminating end can send signals back to the originating end by reversing the battery feed polarity on the ring and tip. The polarity reversal is detected at the originating end by a polarity-sensitive relay or equivalent circuit.

Reverse-battery signaling can be used within SXS offices and on interoffice and PBX trunks. The reverse-battery state is often used to indicate *answer supervision*. Answer supervision indicates that the far-end subscriber has answered the phone.

E&M Lead Signaling

E&M lead signaling is used within an office at the demarcation point between a trunk circuit and a signaling set. The E and M leads do not connect to interoffice cable pairs. Two types of E&M signaling are described below, and Reference 13 covers three additional types.

Type I. The traditional E&M lead scheme (now called *type I*) is shown in Fig. 1-11a. The trunk circuit signals off-hook to the signaling set by transferring the M lead from ground to battery (i.e., -48 V). Resistor R_1 is a current-limiting device such as a resistance-lamp or positive temperature coefficient thermister. Resistor R_2 is a surge-limiting device such as a 1000-Ω power resistor or, in newer designs, a 65-V zener diode. The signaling set signals off-hook to the trunk circuit by grounding the E lead.

The two directions of signaling each have their own conductor and are independent. They are also independent from the ring and tip transmission leads.

Type II. Figure 1-11b shows *type II* E&M lead signaling. In this scheme, four wires are used. The M lead is paired with the *Signaling Battery* (SB) lead and the E lead is paired with the *Signaling Ground* (SG) lead. Since it is electrically balanced, type II signaling is less likely than type I to induce noise into other circuits or to cause large ground currents to flow.

Note that two trunk circuits (or two signaling sets) using type II signaling can be wired back-to-back without using an adapter circuit as would be required when connecting two type I circuits back-to-back.

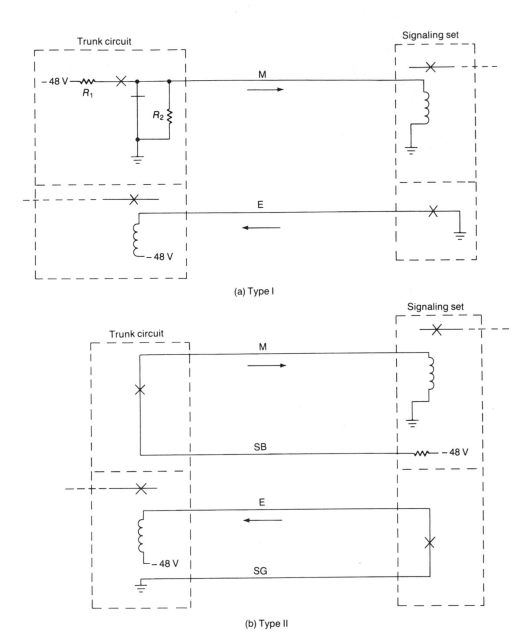

(a) Type I

(b) Type II

FIGURE 1-11   E&M lead signaling. All relays shown on-hook.

Tone Signaling

Tone signaling takes advantage of a circuit's VF transmission path. With signaling in the form of VF tones, there is no need for separate dc signaling leads or the complexity of combining dc signaling and VF transmission on the same wires. Tone signaling is also faster than dc signaling.

Dual-Tone Multifrequency.   *Dual-Tone Multifrequency* (DTMF) is the generic name for the touchtone signaling scheme developed by AT&T. DTMF is used for signaling from a telephone set or PBX to a CO (Fig. 1-12a). After it is seized, but before returning a dial tone, the CO attaches a DTMF receiver to its end of the loop. As each button on the DTMF dial is pressed, the phone's talk circuit is disconnected and a tone pair is sent to the CO. (The tones are intentionally coupled into the talk circuit at a low level to provide user feedback.) The tone frequencies are listed in Table 1-1. DTMF signaling is also used end-to-end (between subscribers).

TABLE 1-1   DUAL-TONE MULTIFREQUENCY SIGNALING

Column Frequency (Hz) / Row Frequency (Hz)	1209	1336	1447	1633[a]
697	1	2	3	A
770	4	5	6	B
852	7	8	9	C
941	*	0	#	D

[a] Special use; not all DTMF dials and receivers can send and receive 1633 Hz.

Multifrequency Key Pulse.   *Multifrequency Key Pulse* (MFKP) signaling is used on interoffice trunks to send called and calling number digits (Fig. 1-12b). We have shown two types of office for variety, analog and digital. When the analog (originating) office is ready to send digits, its common control connects a MFKP *sender* directly to the outgoing trunk circuit. At the digital (terminating) office, the common control connects the incoming trunk circuit to an MFKP *register* via the matrix. After the digits are sent, the sender and register are dropped from the circuit. In MFKP signaling, each digit is sent as a pair of tones—but *not* at the same frequencies as DTMF tones.

Single frequency.   In *Single-Frequency* (SF) signaling the on- and off-hook states of a trunk are sent as the presence and absence of a 2600-Hz tone. Figure 1-12c shows the application of *SF sets* at each end of a four-wire trunk. When the trunk is idle, the SF sets send 2600-Hz tones in each direction over the facility.

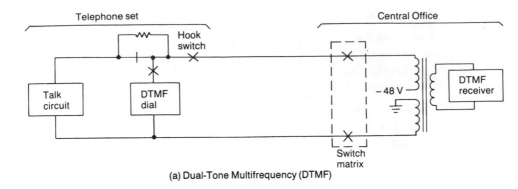

(a) Dual-Tone Multifrequency (DTMF)

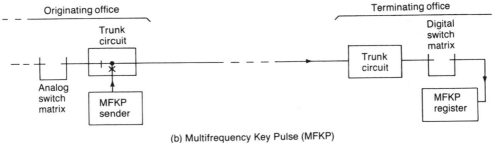

(b) Multifrequency Key Pulse (MFKP)

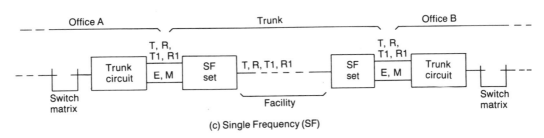

(c) Single Frequency (SF)

FIGURE 1-12　Tone signaling.

Since this is a four-wire facility, there are tip–ring pairs for each direction of transmission. The conductors of one pair are designated T and R; the conductors of the other pair are designated T1 and R1, as shown in Fig. 1-12c.

Each SF set detects the far-end tone and opens the E lead toward the trunk circuit, marking the trunk idle. When either office seizes the trunk, the trunk circuit connects battery voltage to the M lead. The near-end SF set detects the M-lead battery and removes the tone toward the far end. The far-end SF set detects absence-of-tone and grounds the E lead to seize the far-end office. Dial pulses can follow, with each on-hook pulse sent as a burst of 2600-Hz tone. When the distant party answers, the far-end office removes the tone toward the near end to signal answer supervision (off-hook). Single-frequency signaling can be used on two-wire trunks by assigning a different frequency for each direction of transmission.

Ringing

The specific example of ringing a telephone was given at the beginning of this chapter. In the general case, *ringing* is a type of signaling that uses a low-frequency, high-voltage alternating current sent from the originating end of a circuit to seize the terminating end. The power delivered in this scheme can directly operate five electromechanical ringers at the end of a long (e.g., 2000-Ω) subscriber loop. The ac is usually superimposed on a dc voltage known as the *tripping battery*. When the terminating end answers, it closes a dc path around the loop. The tripping battery allows the originating end to detect the dc path and *trip* (or stop) ringing.

Selective ringing is a scheme that manipulates the timing, frequency, polarity, or circuit configuration of the ringing signal in order to alert only one subscriber on a party line. (A *party line* is a loop that is shared by two or more subscribers. Only one subscriber at a time can use its telephone.) Due to selective ringing methods plus the wide range of loop and office conditions, the ringing voltage found on a subscriber loop may vary from 40 to 150 V rms, at 15.3 to 68 Hz, superimposed on -105 to $+75$ V dc. The most common ringing voltage is on the order of 90 V rms, at 20 Hz, superimposed on the nominal -48 V dc office battery. Ringing is used on subscriber loops, loop-start PBX trunks, and private circuits that use *ring down* signaling. Ringing is also found on ground-start PBX trunks, where it can be used in combination with the ground-start signal to alert the PBX attendant.

1-3-4 Transmission and Signaling Applications

Figure 1-13 shows applications of some of the transmission and signaling schemes that we have discussed.

Analog environment. The analog switching matrix in Fig. 1-13a interfaces to the rest of the network via E&M trunk circuits. The standardized demarcation between switching office and trunk plant is located outboard of each trunk circuit.

The trunk on the right is implemented using a cable pair. A *duplex* signaling set interfaces the cable pair to the trunk circuit. (Duplex signaling uses the dc conductivity of the tip and ring conductors plus ground to provide independent signaling paths in each direction over a cable pair.)

The trunk to the left is implemented on a carrier system. This example uses a basic carrier Channel Unit (CU)—one with no built-in signaling or hybrid. Because of the CU's simplicity, two additional blocks of circuitry are needed to interface the carrier to the trunk circuit.

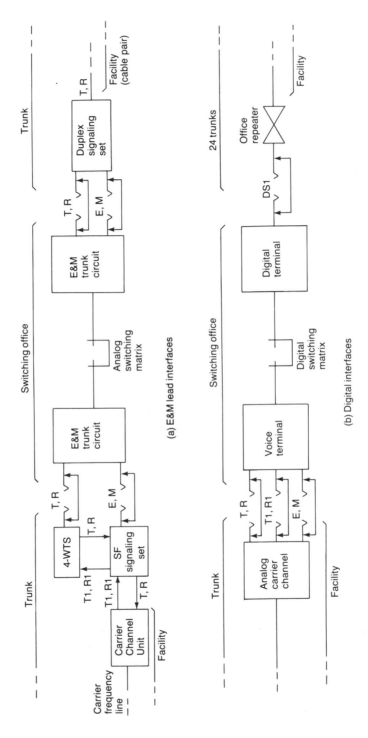

FIGURE 1-13 Transmission and signaling applications.

26

The first block is a *Four-Wire Terminating Set* (4-WTS) which contains a hybrid circuit. The hybrid converts the two-wire circuit from the switch matrix to a four-wire circuit for the carrier channel (carrier is inherently four-wire).

The second block is a single-frequency signaling set. The SF set converts dc signaling on the M lead into 2600-Hz tone signaling on the T and R pair to the CU. The SF set also converts 2600-Hz tone signaling on the T1 and R1 pair from the CU into dc signaling on the E lead to the trunk circuit.

Digital environment. Figure 1-13b shows a digital switching office with both analog and digital transmission facilities attached. To the right, a group of 24 trunks is carried on one digital facility. In this example, the digital facility consists of two wire pairs—one for each direction of transmission. Each pair operates at 1.544 Mb/s. This specific bit rate is designated *DS1* (Digital Signal 1). (Other bit rates in use for digital telephony are designated in the series DS0, DS1, ...) The facility interfaces to the switch via an *office repeater*. This repeater amplifies the digital line signal to a standard amplitude for application to the test jack access. This access point has standardized characteristics and is the demarcation point between the digital switching office and the digital trunk facility. Note that a test connection made at this *DS1* point accesses *24 trunks*.

On the switch side of the demarcation point, we find a *digital terminal*. This terminal converts the switch manufacturer's internal digital signal scheme to a standard DS1 signal.

On the left of the figure, trunks leave the office on an analog carrier system. The interface between trunks and switch is accomplished by equipping the switch with a *voice terminal*. This terminal contains A/D and D/A converters and serves to demultiplex and convert the switch's internal digital signals to standardized VF signals for each trunk.

Compare Fig. 1-13a and b and note that the voice and digital terminals of the digital office are analogous to the trunk circuits of the analog office. In the digital environment, however, the blocks of circuitry serve trunks in multiplexed groups of 24 or more—providing one of the economies of digital switching.

1-4 VOICE TRANSMISSION AND MEASUREMENT

Loop Simulator Circuits

Earlier in this chapter we described *loop signaling*, a dc signaling scheme that is superimposed on the conductors that carry the VF transmission. The dc flowing in a circuit can affect its transmission performance. Many circuits will not function unless dc is present. For these reasons, dc is usually injected into loop-signaling circuits when transmission measurements are made. As a further refinement, the dc amplitude and polarity are made adjustable to simulate a wide range of *loop conditions*.

Figure 1-14 shows a *loop simulator* circuit, whose purpose is to inject dc into a test setup without the simulator itself influencing the transmission performance. This is accomplished by coupling the dc through a large (≥ 10 H) inductor and by coupling the VF through a large (500 μF) capacitor. Switch SW_1 allows the simulator to deliver dc of either polarity. Switch SW_2 allows the simulator to act as either the battery-feed end or the loop-closure end of a loop-signaling circuit. Finally, switch SW_3 allows a choice of VF terminating impedances. Tables 1-2 and 1-3 list the simulator switch settings and component values for various applications.

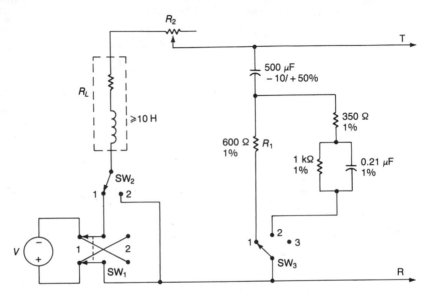

FIGURE 1-14   Loop simulator. (Composite circuit adapted from FCC Rules Part 68. See Tables 1-2 and 1-3 for switch settings and additional component values.)

Example 1-1

In this example we illustrate how the loop simulator would be used in a measurement setup. The unit under test is a piece of loop-start telephone equipment that emits VF tones—perhaps DTMF tones. The controlling specification may read "shall not deliver more than 0 dBm into the loop simulator under all applicable conditions." The "applicable conditions" here are all the switch settings, voltages, and resistances listed in Table 1-2 for loop start circuits.

We wish to check the test specimen first at minimum current. This will occur at minimum battery voltage and maximum simulated loop resistance. Figure 1-15a shows the circuit. So that we can calculate the current, let's assume that the specimen has a resistance of 200 Ω. This yields a current of 21.9 mA dc. Figure 1-15b shows the conditions for the maximum current of 94.2 mA dc.

The transmission parameter under test should be checked over the range of loop currents provided by the simulator, not just at the extremes. Also, do not

TABLE 1-2 LOOP SIMULATOR: DC PORTION[a]

Circuit type	Switch settings		V (V)	$R_2 + R_L$ (Ω)		
	SW_1	SW_2				
Loop and ground start	1 and 2	1	42.5–56.5[b]	400–1740		
	2	1	105	1300–2000[c]		
Reverse battery	N/A[d]	2	N/A	400–2450		
Off-premises station[f]				Class A[e]	Class B	Class C
	N/A	2	N/A	R_L–200	R_L–800	R_L–1800
	1	1	24	N/A	200–2300	900–3300

[a] Composite circuit values adapted from Part 68 of FCC Rules[1] and EIA-464 (PBX)[4].

[b] FCC Part 68 relaxes the upper limit to 52.5 V. We recommend testing at the higher voltage.

[c] FCC Part 68 allows testing at 2000 Ω only.

[d] Not applicable.

[e] The *class* of an off-premises station circuit refers to its operating loop resistance limit.

[f] For Off-Premises Station (OPS) circuit testing, $R_L \ll$ maximum R_2. Under all test conditions, the OPS circuit must deliver 16 mA dc minimum to the loop simulator. The OPS circuit battery feed polarity is negative on ring.

assume that the resistance of a terminal device is linear with current, as in the example.

Instruments designed for telephone applications usually contain built-in *hold coils* or electronic *hold circuits*. Often simply an inductor, these circuits provide a relatively low (e.g., 200-Ω) dc resistance and high ($\geqslant$900-Ω) ac impedance. Thus they *hold* a loop-signaling circuit by passing dc, yet do not shunt voice frequencies. Some instruments also provide a built-in battery-feed circuit, also with low dc resistance and high VF impedance. You may be tempted to use these built-in circuits instead of the external loop simulator circuit. Use of built-in dc circuits may cause errors, however. In one case, the built-in inductor or equivalent circuit may saturate at high currents and lower the VF impedance of the instrument. In another case, there may be a small-valued built-in inductor. This would cause the instrument's low-frequency response to suffer when its *hold* circuit is switched in.

TABLE 1-3 LOOP SIMULATOR: AC PORTION

Condition	Switch setting (SW_3)
Normal	1
Alternative termination for measuring signal power	2
Measuring longitudinal balance	3

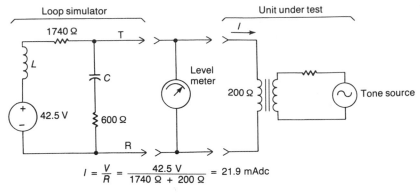

$$I = \frac{V}{R} = \frac{42.5 \text{ V}}{1740 \ \Omega + 200 \ \Omega} = 21.9 \text{ mAdc}$$

(a) Case I: Minimum direct current

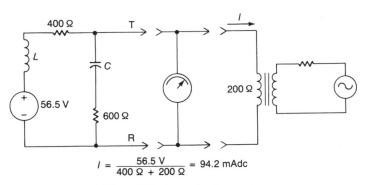

$$I = \frac{56.5 \text{ V}}{400 \ \Omega + 200 \ \Omega} = 94.2 \text{ mAdc}$$

(b) Case II: Maximum direct current

FIGURE 1-15   Application of loop simulator.

Test Instruments

The instruments used for measuring the transmission characteristics of telephone circuits have been especially adapted for use in the telephone environment.
As discussed in the preceding section, these instruments contain dc paths (hold
circuits) that may be switched in to hold a working circuit during test. Other
features include balanced inputs and outputs, selectable load and source impedances (600 and 900 Ω are standard at voice frequencies), and a variety of connec-
tor types.

Test connectors.   The traditional VF telephone connectors for central office use are the *310 plug* and mating *jack*. A 310 plug is about the same size and

configuration as the $\frac{1}{4}$-inch consumer audio *phone plug*, which is derived from the 310 plug. The 310 plug is, however, a precision, machined part designed for thousands of trouble-free insertions over its lifetime. (Think of the heavy use this plug would get in a manual *cord switchboard*.) Since the dimensional outline of a 310 plug differs from that of a phone plug, we advise against mixing the two. Loose connections and damage to the jack could occur. The *bantam* plug and jack are newer, smaller replacements for the 310 plug and jack. Bantam connectors have no counterpart in consumer audio.

Each input and output of an ideal test instrument would have parallel-connected jacks for 310 and bantam plugs (for telephone use) plus a set of banana jacks (for general lab use).

Throughout the book, we illustrate test instruments designed for various applications, with different degrees of internal automation and from an assortment of manufacturers. The small sample provided merely shows the variety available. In almost all cases, alternative instruments are available from other manufacturers.

Transmission Characteristics

This chapter has provided an overview of both signaling and transmission. The remainder of the book covers voice-frequency transmission in more detail. Chapters 2 through 7 follow a common format for each topic: introduction of concepts, measurement or design examples, review of industry performance, and presentation of published industry standards. The standards are grouped in tables at the end of each chapter for easy access. The arrangement of these tables by transmission topic instead of by source document creates a unique reference.

Levels, loss, and frequency response (Chapter 2).   Understanding the topic of transmission levels or amplitudes is the key to benefitting from later chapters. Almost all transmission measurements are related in some way to a knowledge of signal levels. Adequate frequency response is important for both voice intelligibility and voice-band data reliability.

Noise and crosstalk (Chapter 3).   We define *noise* as any unwanted signal found in a channel, in the presence or absence of the desired signal. *Crosstalk* is the desired signal from one channel found in another channel, where it is not desired.

Hybrids (Chapter 4).   *Hybrids* are circuits that convert two-wire circuits to four-wire circuits. This is, hybrids split the two directions of transmission (on a two-wire circuit) into separate directions of transmission (on a four-wire circuit).

Return loss and echo (Chapter 5).   *Echo* occurs in the telephone network when a signal is reflected from an impedance discontinuity—often at a hybrid. *Return loss* is a measure of the severity of the reflection.

Longitudinal balance (Chapter 6). Longitudinal balance is a measure of the coupling between normal-mode and common-mode signals on balanced circuits. Poor longitudinal balance can result in noise and crosstalk.

Distortion and miscellaneous impairments (Chapter 7). Distortion exists only in the presence of a signal. Nonlinear distortion creates unwanted signals at frequencies other than those of the desired signal. Linear distortion merely changes the desired signal without creating other frequencies. The *miscellaneous impairments* of Chapter 7 are those that affect voice-band data more than voice. Examples are hits, dropout, and jitter.

Standards Organizations

Chapter 8 lists the standards cited in the text by their issuing organization. Prices and ordering information are provided. Chapter 8 also lists several additional references.

2
LEVELS, LOSS, and FREQUENCY RESPONSE

The transmission *level* of a telephone signal refers to its amplitude. This could be a voltage or current amplitude, or the amplitude of the power delivered to a load. Amplitudes can be expressed as peak, peak-to-peak, or rms. Telephone signal levels are usually stated in terms of *rms power*.

Transmission levels can be absolute or relative, resulting in a variety of units for expressing signal levels. Absolute power is measured in units of watts, dBm, and dBrn. Relative power is measured in dB, dBm0, and dBrn0. In some cases, voltage and current ratios can be expressed in dB. Frequency-weighted power measurements call for additional units such as the dBrnC and the dBrnC0 (pronounced *dee-brink-oh*).

In this chapter we present the concepts of voice-frequency (VF) telephone transmission levels and their units of measurement. The emphasis is on a progressive development starting from one basic formula. If you follow this development, you should be able to recreate, at any time, the theory of transmission levels. You will not have to rely on recall of dB tables or formulas (except the first one). To present the development in a logical sequence, we may deviate somewhat from history.

In addition to levels, the topics of this chapter include impedance, transmission level point, loss plans, level stability, level tracking, frequency response, gain, loss, and level-measuring instruments.

2-1 LEVELS AND LOSS

2-1-1 Relative Levels

Decibels, Gain, and Loss

The bel.   We start with the *bel*, a unit named after Alexander G. Bell. The bel expresses the ratio between two powers, P_1 and P_2:

$$\text{power ratio (in bels)} = \log \frac{P_1}{P_2} \tag{2-1}$$

[In this book, *log* means *common log* (i.e., log base 10). Also, unless specified otherwise, power is *rms power*.]

The decibel.   Instead of the bel, we usually use a smaller unit, the *decibel* (dB):

$$\text{power ratio (in dB)} = 10 \log \frac{P_1}{P_2} \tag{2-2}$$

The two powers whose ratio we want may appear at two different points of a circuit, as in the following example.

Example 2-1

A phonograph cartridge delivers 2 μW to a stereo amplifier, and the amplifier boosts this signal and delivers 1 W to a loud speaker. This is a power ratio of 1 W/2 μW = 500,000. Expressed in decibels, the power ratio is

$$10 \log \frac{1 \text{ W}}{2 \text{ }\mu\text{W}} = 57 \text{ dB}$$

Gain.   When the power increases, we have a *gain* and the power ratio expressed in dB is positive. We would say there is a 57-dB *power gain* through the amplifier.

When we express a *power* ratio in decibels, the impedances at the two circuit points need not be equal. In our example of phono input and speaker output, the impedances are definitely not equal.

Two signal powers of interest may appear at the same circuit point but under different conditions, as in the next example.

Example 2-2

Let's "turn down" the amplifier in Example 2-1 so that we are delivering 200 mW to the speaker instead of 1 W. This change represents a power ratio of

$$\frac{200 \text{ mW}}{1 \text{ W}} = 0.2$$

or

$$10 \log \frac{200 \text{ mW}}{1 \text{ W}} = -7 \text{ dB}$$

Loss.   When the power decreases, we have a *loss* and the power ratio expressed in dB is negative.

Adding and subtracting decibels.   Let's examine one of the great benefits of expressing power ratios in decibels.

Example 2-3

Recall that the amplifier in our first example had a gain of 57 dB. Then we turned down the amplifier (decreased the gain) by 7 dB. What is the amplifier's new gain? It is simply 57 dB − 7 dB = 50 dB. [To verify this, note that $10 \log (200 \text{ mW}/2 \, \mu\text{W}) = 50$ dB.]

In the example we have used a mathematical tool: The product of two numbers is equal to the inverse log of the algebraic sum of their logarithms. Note that we mention *algebraic sum*. This is to alert you to be careful with signs when adding and subtracting decibels. For illustration, we can consider our action of turning down the volume control as introducing either a 7-dB loss or a −7-dB gain. To avoid errors when combining gains and losses, you may want to be rigorous with the math by following these rules:

1. Convert all gains or losses to gains.
2. Take the sum of the gains.
3. Convert the sum to a loss if desired.

Using these rules for Example 2-3, we obtain the following:

1. Amplifier gain = 57 dB
   Volume control gain = −7 dB
2. Sum = 57 dB + (−7 dB) = 50 dB

Voltage ratios.   When dealing with signals in electronic circuits, it is often more common to express amplitudes in volts than in watts. Although defined for power ratios, the decibel can still be a useful tool when voltage is specified. The straightforward approach is to convert the voltages to power and then use Eq. (2-2). However, if appropriate formulas are developed, a *voltage ratio* can be used to find a *power ratio* expressed in decibels. First, we examine the simplifying case where the two voltages in question, V_1 and V_2, appear across *equal* impedances.

Since $P = V^2/R$, Eq. (2-2) becomes

$$10 \log \frac{P_1}{P_2} = 10 \log \frac{V_1^2/R}{V_2^2/R}$$

$$= 10 \log \left(\frac{V_1}{V_2}\right)^2$$

$$= 20 \log \frac{V_1}{V_2}$$

That is,

$$\text{power ratio (in dB)} = 20 \log \frac{V_1}{V_2} \qquad (2\text{-}3)$$

where the impedances at V_1 and V_2 are equal. Note that V_1 and V_2 may be expressed in volts rms, volts peak, and so on, but they must be in the same units.

We now examine the general case where the impedances at V_1 and V_2 are *not equal*. Substituting for P_1 with V_1^2/R_1 and for P_2 with V_2^2/R_2, Eq. (2-2) becomes

$$10 \log \frac{V_1^2/R_1}{V_2^2/R_2} = 10 \log \frac{V_1^2}{R_1} - 10 \log \frac{V_2^2}{R_2}$$

$$= 10 \log V_1^2 - 10 \log R_1 - 10 \log V_2^2 + 10 \log R_2$$

$$= 20 \log \frac{V_1}{V_2} - 10 \log \frac{R_1}{R_2}$$

That is,

$$\text{power ratio (in dB)} = 20 \log \frac{V_1}{V_2} - 10 \log \frac{R_1}{R_2} \qquad (2\text{-}4)$$

where the impedances at V_1 and V_2 are not equal.

Table 2-1 lists a wide range of decibel values and their corresponding power and voltage ratios. By studying the table, you will note some rules of thumb well known to transmission engineers. For example, a gain of 3 dB corresponds to a doubling of power; a 6-dB loss halves the voltage; and so on.

2-1-2 Absolute Levels

The dBm. The decibel is a unit used to express the *ratio* of two powers. We can use the decibel to say that one signal is so many dB *relative* to another signal. The unit *dB* does not tell us the *absolute* value of either signal.

Absolute power is expressed in watts or in another absolute unit, the *dBm*. The unit dBm means "decibels relative to one milliwatt." Thus 0 dBm is a power of 1 mW, and 20 dBm is a power 20 dB higher than 1 mW. So 20 dBm equals 100 mW [use Table 2-1 or Eq. (2-2) to see this]. Similarly, −20 dBm equals 0.01 mW.

TABLE 2-1 DECIBELS

Decibels	Power ratio	Voltage ratio
−90	10^{-9}	3.16×10^{-5}
−60	10^{-6}	0.001
−50	10^{-5}	0.00316
−40	10^{-4}	0.01
−30	0.001	0.0316
−20	0.01	0.1
−10	0.1	0.316
−6	0.251	0.501
−3	0.501	0.708
−2	0.631	0.794
−1	0.794	0.891
−0.5	0.891	0.944
−0.1	0.977	0.989
0	1.000	1.000
0.1	1.023	1.012
0.5	1.122	1.059
1	1.259	1.122
2	1.585	1.259
3	1.995	1.413
6	3.981	1.995
10	10	3.16
20	100	10
30	1000	31.6
40	10^4	100
50	10^5	316
60	10^6	1000
90	10^9	3.16×10^4

Significance of impedance.   If we know the circuit impedance, the power also tells us the voltage:

$$V = (PR)^{1/2} \tag{2-5}$$

where P is in mW, R is in kΩ, and V is in volts; or

$$V = R^{1/2} \times 10^{P/20} \tag{2-6}$$

where P is in dBm, R is in kΩ, and V is in volts. In cases where we assume rms power in the two previous equations, V is in *volts rms*.

An impedance of 600 Ω (resistive) is common for terminating VF circuits. A signal power of 0 dBm into 600 Ω is 0.775 V rms [from Eq. (2-6)]. Another common impedance is 900 Ω. A level of 0 dBm into 900 Ω is 0.949 V rms. If you work frequently in this field, these two voltages will become quite familiar. If they do not, they are easily derived. For example, when considering 0 dBm (600 Ω), just ask: What voltage across 600 Ω dissipates 1 mW?

In telephone work, it is common practice to use terminated circuits and to match impedances. In Fig. 2-1 we show a signal source on the left driving a load (R_L) on the right. The signal source comprises a voltage source V_1 *and* a source resistance R_S. For impedance matching, $R_L = R_S$. The source delivers maximum power to R_L when $R_L = R_S$. This circuit is said to be "terminated in 600 Ω."

2-1-3 Instruments for Measuring Levels

A transmission *level meter* is simply a voltmeter that has its scale marked in units of dBm instead of volts. Even though the dBm is a unit of power, a level meter does not measure power directly. Instead, it measures the voltage across a terminating resistor. The resistor's value is equal to the circuit's nominal impedance, usually 600 or 900 Ω for VF circuits. The resistor may be internal or external to the meter. Since the meter measures a voltage but indicates in units of power, the scale calibration must be based on the value of the terminating resistor.

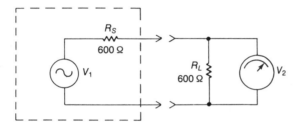

FIGURE 2-1 Impedance matching.

Meter Response and Calibration

Alternating-current voltmeters can be designed with a variety of response characteristics. An *average-responding* meter measures the average value of a signal after it has been rectified. *Peak-responding* and *rms-responding* meters measure the peak and rms values of the input waveform. Regardless of the response characteristics, voltmeter scales are typically calibrated using the units *rms volts* or *peak volts*.

A common design used in level meters is the *average-responding/rms-calibrated* meter. This instrument measures the average value (after rectification) of the input waveform, converts the average value to an rms value, and displays the result using the unit *rms volts*. The conversion factor is based on a sine-wave input. In general, this instrument will not indicate the correct voltage unless the input is a sine wave. Since telephone transmission level measurements are made using sine waves, average-responding/rms-calibrated meters are suited for this application.

On the other hand, noise and distortion waveforms are usually complex, not sinusoidal. Noise and distortion should be measured using an rms-responding meter. These instruments are often called *true rms* to distinguish them from average-responding meters.

A *quasi-rms-responding* meter may also be used for noise and distortion measurements. Quasi-rms is loosely defined as a response that is sufficiently close to true rms that a waveform of interest is measured with minimal error. Standards that allow quasi-rms response (such as IEEE 743-1984) contain more quantitative definitions.

Figure 2-2 shows a typical *Transmission Measuring Set* (TMS). This set measures level and frequency, and it contains a test oscillator. It also measures noise and quantizing distortion, subjects discussed later. Figure 2-3 is a simplified schematic of a level meter or of the level-measuring portion of a TMS.

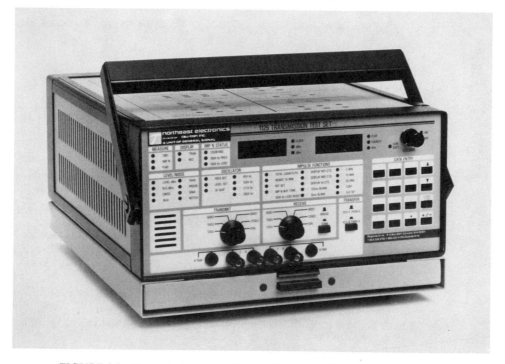

FIGURE 2-2   Transmission measuring set. (Courtesy of Northeast Electronics.)

Terminating versus Bridging Measurements

Recall that telephone circuits are usually designed to work into a termination. If the termination is present in the setup being tested (i.e., external to the meter), the level meter is set on *bridging* (see switch SW_1 in Fig. 2-3). If the level meter itself is to provide the circuit termination, the meter is set on *terminating*.

A common measurement error is to *double terminate* a circuit by setting the instrument on *terminating* when measuring an externally terminated circuit.

For both terminating and bridging measurements, switch SW_2 selects the impedance. The dotted line from SW_2 into the voltmeter reminds us that the voltage-to-power conversion factor is a function of impedance.

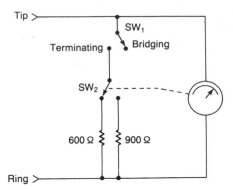

FIGURE 2-3   Simplified level meter.

2-1-4 Design Examples

We have now presented enough of the theory of transmission levels to use the information in a circuit design.

Example 2-4

Design a simple telephone amplifier with the following specifications: based on an operational amplifier (op amp), 600-Ω input and output impedances, and 8-dB gain. You may make the following assumptions and simplifications: ideal op amp, single-ended (unbalanced), and no dc isolation of input or output.

Figure 2-4 shows the circuit. In the figure we also show the source and load that would typically be connected to the amplifier. Resistors R_1 and R_4 are set at 600 Ω to provide the specified input and output impedances. R_2 is set at 100 kΩ so as not to load R_1. R_3 now determines the amplifier's gain.

We were given a gain specification of 8 dB. Implied in the specification is that this is a *power gain*. The amplifier should deliver 8 dB more power to R_L than the source delivers to R_1. We will convert the 8-dB power gain to a voltage ratio between

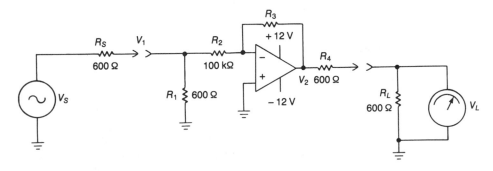

FIGURE 2-4   Amplifier with 8-dB gain.

V_1 and V_L. Since R_1 and R_L are equal impedances, we can use Eq. (2-3):

$$8 \text{ dB} = 20 \log \frac{V_L}{V_1}$$

$$\frac{V_L}{V_1} = 2.51$$

(2-7)

Note that $V_L = V_2/2$ (since $R_4 = R_L$). Substituting in Eq. (2-7) gives

$$\frac{V_2/2}{V_1} = 2.51 \qquad \text{or} \qquad \frac{V_2}{V_1} = 5.02$$

For our inverting op-amp stage,

$$\frac{V_2}{V_1} = \frac{R_3}{R_2} = 5.02$$

Now

$$R_3 = (5.02)R_2 = 5.02 \times 100 \text{ k}\Omega = 502 \text{ k}\Omega$$

We will make $R_3 = 499$ kΩ, which is the nearest standard 1% resistor value. This completes the design.

Example 2-5

What is the maximum output level (in dBm) that the amplifier in Example 2-4 can deliver to its load?

Assume that the op amp's output is specified to swing to within 2 V of the supply rails when driving the 1200-Ω load presented.

With a ± 12-V supply, the op amp will deliver 20 V peak-to-peak at V_2. This is 7.07 V rms. The voltage drops in half at V_L, to 3.54 V. We know that 0 dBm into 600 Ω is 0.775 V [Eq. (2-6)], so 3.54 V into 600 Ω is

$$20 \log \frac{3.54 \text{ V}}{0.775 \text{ V}} = 13.2 \text{ dBm}$$

2-1-5 Impedance Mismatch

As we will see in Chapter 4, an impedance mismatch can have a substantial negative effect on return loss and trans-hybrid loss. However, the effect on level measurements is usually small. Consider the case of measuring the output level of a nominal 600-Ω circuit (Fig. 2-5a). For convenience, we will let the level be 0 dBm. The voltages are then as shown.

Now let us set the level meter on 900 Ω in error (Fig. 2-5b). This presents a higher impedance (less load), so the voltage at the meter input increases. However, we are measuring power, not voltage, and 0.930 V into 900 Ω is -0.2 dBm. This fractional-dB error is not significant in many cases. The converse mistake (measuring a 900-Ω circuit with the meter set on 600 Ω) also yields a -0.2-dB error.

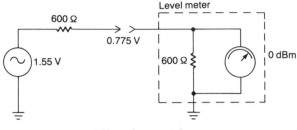

(a) Impedance match

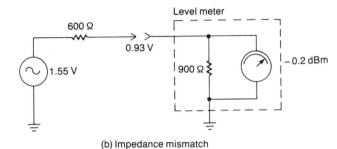

(b) Impedance mismatch

FIGURE 2-5 Level measurement.

Note that the discussion above applies to a meter set on *terminating*. If a meter is used in its bridging mode and the impedance selector control is set incorrectly, the error is usually significant. Bridging a 600-Ω circuit with a *900-Ω* meter will result in a −1.8-dB level measurement error. Bridging a 900-Ω circuit with a *600-Ω* meter will result in a +1.8-dB level measurement error. You may wish to test your knowledge of transmission level measurement by confirming these results mathematically.

Insertion Loss versus Transducer Loss

The terms *insertion loss* and *transducer loss* both appear in the standards literature. *Insertion loss* (in dB) is defined as $10 \log (P_1/P_2)$, where P_1 is the power delivered from the source directly to the load with the circuit under test removed, and P_2 is the power delivered to the load with the circuit under test inserted.

Transducer loss (in dB) is defined as $10 \log (P_3/P_2)$, where P_3 is the maximum power available from the source, and P_2 is defined above. (Recall that the maximum power available from a source is the power that the source would deliver to a load resistance equal to the source resistance.)

When the impedances of the test oscillator and level meter are set to the same value, insertion loss equals transducer loss. In general, test instrument impedances are set to match the port to which they attach and not to match each other. Thus transducer loss is measured. In this book, *loss* usually means *transducer loss*.

With a 600-Ω instrument at one end of a circuit and a 900-Ω instrument at the other end, the difference between transducer loss and insertion loss is only 0.2 dB.

2-1-6 Transmission Pads

Transmission pads (or simply *pads*) are attenuators. In telephony, pads are used to reduce standardized equipment levels to the specific levels desired for various applications. Examples of pad use in trunk design are presented later in this chapter.

Pads can be classified in several ways. There are balanced pads, unbalanced pads, impedance-matching pads, and pads that do not match impedances. There are a number of circuit configurations, such as the *L-pad, T-pad, H-pad,* and *O-pad.* In this book we treat pads as black boxes—specifying only their loss. You may assume that inside the box is a balanced circuit (probably an O-pad) that matches the applicable circuit impedance at both its input and output.

2-1-7 Dynamic Range

The concept of dynamic range is the same in telephony as it is in other audio fields, such as broadcasting and hi-fi. At any point in a circuit there exists a useful range of signal amplitudes. Signal amplitudes above this range become distorted due to circuit overload. Signals at levels below this range are lost in noise. This range of useful amplitudes is the *dynamic range*.

Our purpose in introducing dynamic range is to provide a concept useful in discussions that follow. In telephony, dynamic range is usually not specified directly or measured as a routine.

Test Tone Level

Test tone level (or simply *test level*) is a reference signal located in amplitude near the high end of a circuit's dynamic range. In fact, the test tone level serves to anchor the dynamic range in the continuum of absolute signal amplitudes. The endpoints of the dynamic range are measured relative to test tone level. The test tone is a 1004-Hz sine wave.

To illustrate these concepts, we consider an intertandem trunk in Fig. 2-6a. The trunk connects two tandem switching offices. Our measurement point is at office A, and we are measuring signals as they leave office A bound for office B. A frequent test made on trunks involves sending a test tone from one end to the other. This tests the trunk's VF functioning and allows its loss to be measured. The test tone is sent at *test tone level,* −3 dBm in our example.

Note that test level is a fairly high amplitude signal. (Talker voice levels are typically 15 to 20 dB below test level.) To minimize crosstalk or other interference

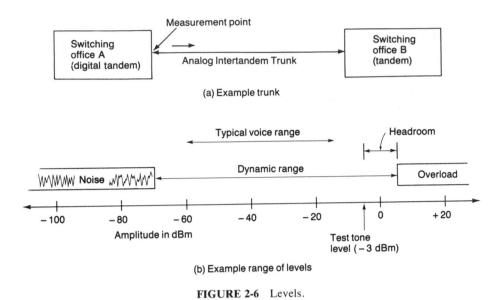

FIGURE 2-6 Levels.

from test tones, testing and alignment of in-service equipment are often done
using tones that are 10 dB below test level. This technique is especially likely to be
used when tones must be applied for more than a few seconds or when multiple
channels may simultaneously be excited with steady tones.

Overload

At some amplitude above test tone level, the circuitry used to implement the
transmission circuit may saturate or *overload*. Signals at levels higher than the
overload point are significantly distorted. The difference between test tone level
and the overload point is the *headroom.*

For trunks, the overload point is a function of the facilities that make up the
trunk. [A *facility* is a transmission medium such as a wire pair or a carrier (multiplex) system. Two or more facility types may be connected together to implement
a single trunk.]

For an example of overload point, a PCM carrier system has a rather sharp
overload point located 3 dB above test tone level. (The overload point is sharp in
digital systems since the PCM encoder runs out of higher-amplitude bit patterns at
overload.) At the other extreme, a simple nonrepeated wire pair has virtually no
overload point. (In practice, however, you cannot apply an arbitrarily high level to
a wire pair due to crosstalk consideration.)

Overload is often specified by giving the signal compression characteristic at
levels near the overload point. Overload objectives are listed in Table 2-4 at the
end of the chapter.

The typical voice amplitude range lies within the dynamic range, as shown in Fig. 2-6b. An occasional loud talker may produce peaks approaching overload, and some soft talkers may get lost in the noise.

2-1-8 Transmission Level Point

We have been discussing the testing of transmission circuits by inserting a test tone at the circuit's input and measuring the resulting output. The test tone is inserted at an amplitude we called the *test tone level*. Test tone level is within and near the top of the dynamic range of the circuit under test. How is the test level selected? Before we can answer, we just first define and discuss a new term.

A *Transmission Level Point* (TLP) is a reference point along a transmission path or circuit. The relative level at any other point in the circuit can be determined if the net gain between the two points is known. The TLP may be an analog point at a standard impedance (directly accessible by a test instrument), or it may be a digital point (where the bit stream must be decoded before measurement).

During maintenance, a test tone is inserted at the TLP. When the test tone is measured at other points downstream in the circuit, its level will be a function of the gains and losses between the TLP and the downstream point.

A transmission level point is just that—a *point* along a circuit. A TLP is further defined by the test tone level that is applied at that point during maintenance. A *0-dBm transmission level point* (often shortened to *0 TLP*) is a point where the applied test level is 0 dBm. Transmission level points have been assigned standard levels and locations in the telephone network.

Figure 2-7 shows that for switching offices the TLP is located at the *outgoing side* of the switch. The significance of this follows.

Consider a built-up connection that passes through a switching office. As the signal passes through the office, it is attenuated by wiring, coupling transformers, and coupling capacitors. The losses of these elements add together to make up the *office loss* (usually a fraction of a decibel). The level measured at the outgoing side of the office will differ from the level measured at the incoming side. Consistent use of the outgoing side for the TLP assures that office loss will be included in trunk loss measurements. Both the subscriber end and the local CO end of subscriber loops are considered to be 0 TLP (Fig. 2-8).

Four-wire carrier systems have standardized levels at their input and output terminals (Fig. 2-9). The inputs operate at a -16 TLP and the outputs at a $+7$ TLP. Viewed on a stand-alone basis, a four-wire carrier channel thus provides 23 dB of gain. As we shall see, such a channel is combined with other elements, such as hybrids and pads, to make up a trunk that has net loss.

The dBm0

The unit *dBm0* is defined as dBm referenced to the zero transmission level point. For example, -10 dBm0 at a $+7$ TLP is a power of -3 dBm. Note that the

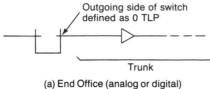

(a) End Office (analog or digital)

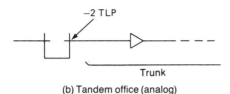

(b) Tandem office (analog)

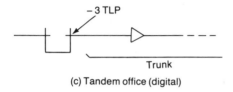

(c) Tandem office (digital)

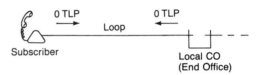

FIGURE 2-7 Transmission level points of switching offices.

FIGURE 2-8 Transmission level point of subscriber loop.

units dB and dBm0 express *relative* powers. The unit dBm expresses an *absolute* power.

Example 2-6

A carrier system specification calls for measuring distortion at -13 dBm0. What absolute power level do we apply to the input of a four-wire channel when making this measurement?

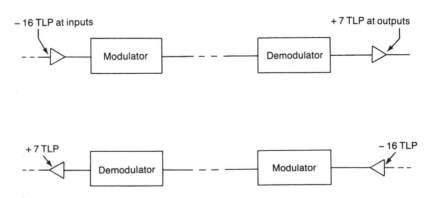

FIGURE 2-9 Transmission level points of four-wire carrier system.

Four-wire carrier channel inputs are at a -16 dB TLP, so apply a level of -13 dBm0 $-$ 16 dB $= -29$ dBm.

2-1-9 Level Tracking

The output level of a linear telephone channel or device should track the input level. This ability is known as *level tracking*. If the input level drops 10 dB, for example, the output should also drop 10 dB. Level tracking is usually specified relative to 0 dBm0 (i.e., test tone level). For example, Bellcore specifies that digital channels have a level-tracking error of <0.5 dB over an input range of -37 to $+3$ dBm0. Full level-tracking objectives are listed in Table 2-5 at the end of the chapter.

Example 2-7

Figure 2-10 illustrates the test setup for measuring level tracking on a four-wire digital channel. We have shown hypothetical data over the full measurement range of $+3$ to -50 dBm0. Measurement step (a) establishes a reference for the rest of the measurements. In (a), an absolute level of -16 dBm (corresponding to a relative level of 0 dBm0) is applied to the channel input. At the channel output, we measure

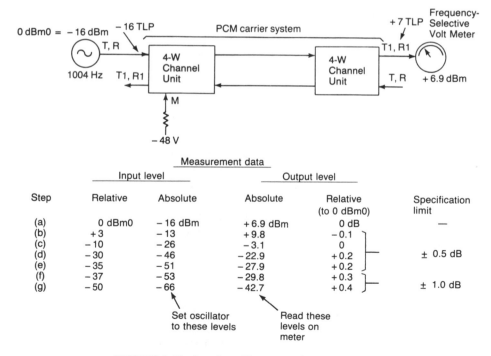

Step	Input level		Output level		Specification limit
	Relative	Absolute	Absolute	Relative (to 0 dBm0)	
(a)	0 dBm0	-16 dBm	$+6.9$ dBm	0 dB	—
(b)	$+3$	-13	$+9.8$	-0.1	
(c)	-10	-26	-3.1	0	
(d)	-30	-46	-22.9	$+0.2$	$\pm$ 0.5 dB
(e)	-35	-51	-27.9	$+0.2$	
(f)	-37	-53	-29.8	$+0.3$	
(g)	-50	-66	-42.7	$+0.4$	$\pm$ 1.0 dB

FIGURE 2-10 Level tracking: example measurement.

an absolute level of +6.9 dBm, which is 0.1 dB below the expected output of +7 dBm. We compensate by considering other output levels to be *relative to +6.9 dBm*.

At step (b), the input is raised 3 dB, to −13 dBm. For linear tracking, the output is also expected to rise 3 dB, from +6.9 dBm to +9.9 dBm. Instead, we measure +9.8 dBm at the output. This is 0.1 dB too low. Thus the output at +3 dBm0 is −0.1 dB *relative* to the output at 0 dBm0, as entered in the table. The tracking specification allows a ±0.5-dB error at +3 dBm0, so data point (b) is within specifications. When the remaining steps are similarly measured, they are all found to be within specified limits.

In the preceding example, two measurement techniques are used to assure accuracy at low levels. First, the channel under test is put in its normal, off-hook state by properly conditioning its M signaling lead. (The state of the signaling bit can affect levels in some systems.) Second, a Frequency-Selective Voltmeter (FSVM) is used at the receiving end to reduce the effects of noise on low-level signals. (A FSVM measures signal power in a narrow band to which it is tuned.)

The level tracking of other circuit types (two-wire channels and switching offices, for example) is measured similarly to the four-wire channel of the example. Test levels must be adjusted, of course, for the various TLPs.

2-1-10 Level Stability

Level stability refers to the long-term accuracy of the loss of a channel or the output level of a piece of equipment. *Long-term* here is in terms of years. Level stability requirements are met at the equipment design stage by proper design and attention to the drift characteristics of electronic components. Table 2-6 lists level stability objectives.

2-1-11 Frequencies Used for Level Measurements

Traditionally, 1000 Hz has been the standard frequency for making measurements of gain, loss, and level. A data point at 1000 Hz is also the reference for frequency response measurements. Now, levels in voice-frequency systems are usually measured at a frequency that is slightly removed from 1000 Hz. Frequencies very near 1000 Hz are avoided since they may beat with the 8-kHz sampling rate of digital voice systems. The beat would cause a wavering of the level measurement and the introduction of unwanted frequency components. Typical test frequencies are thus 1004, 1010, or 1020 Hz.

Note that the *Digital Milliwatt* (DMW) found in digital voice equipment is made to be exactly 1000 Hz by definition of the digital code that produces it. A system's DMW is used to align that system's D/A converters. The DMW should not be decoded and used as an analog signal to test other digital systems since its frequency is exactly (or almost exactly) 1000 Hz.

If a stored digital representation of a 0-dBm0 *1004* Hz sinusoid is used as a test signal, it is called a *Digital Reference Signal* (DRS). To review, the DMW is 1000 Hz, the DRS is 1004 Hz, and both have a level of 0 dBm0.

2-2 TRANSMISSION PLANS

The trunks and switching offices of a telephone network are physically arranged according to a plan. Superimposed on that arrangement is the network's *transmission plan*. The transmission plan specifies the levels, losses, and other voice-frequency characteristics of the network's physical plant.

In the United States the transmission plan of the network has been evolving for over 100 years. The factors influencing this evolution include economics, maintenance, performance, regulation, compatibility of the new with the old, and available technology. In this section we do not simply state a transmission plan for your acceptance. Rather, we present a development of a transmission plan. Our purpose is to allow you to recreate the plan (or some other plan) based along the lines of our development. Note that our presentation is a *logical* development and not necessarily an accurate *historical* development. The telephone network for which our transmission plan is developed is a simplified version of the U.S. network.

2-2-1 Offices and Trunks

Our simplified telephone network comprises two types of switching offices interconnected by trunks (Fig. 2-11). Subscriber's telephones connect to an *End Office* (EO) via local loops. End offices in turn connect to *tandem offices* via *Tandem-Connecting Trunks* (TCTs). Tandem offices interconnect via Intertandem Trunks (ITTs).

There are only two call types possible in our simplified network: local calls and toll calls. A local call connects two telephones within the same EO: for example, telephone A to telephone B. A toll call connects two telephones in different end offices. A minimum-distance toll call passes through two local loops, two end offices, two tandem-connecting trunks, and one tandem office (for example, telephone C to D). A longer-distance toll call may also pass through *two* tandem offices and an intertandem trunk (for example, telephone A to telephone D). Even longer-distance toll calls may pass through additional tandem offices and intertandem trunks. In a practical network, a limit is set on the maximum number of intertandem trunks used to complete a call. (In this discussion, *distance* refers to electrical distance, not airline distance. Because of such considerations as service boundaries, call routing, traffic patterns, and geographical features, the electrical or transmission distance of a connection may relate only roughly to its airline distance.)

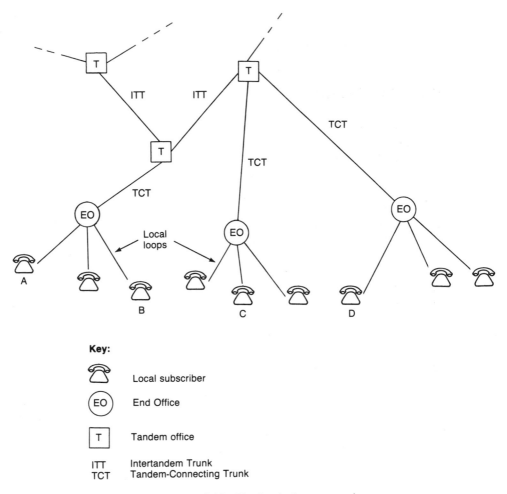

FIGURE 2-11 Simple telephone network.

So far we have developed a network that allows a variable number of trunks to be connected in tandem to complete a call between two telephones. The ability of the two callers to hear each other will in part depend on the loss between their telephones. This loss is the sum of the losses of all the loops, trunks, and offices involved in the call. We now continue with the development of our transmission plan.

2-2-2 Telephone Sets

From a transmission perspective, a telephone set consists of electroacoustical transducers. The *mouthpiece,* or *transmitter,* converts acoustical energy into electrical energy. The traditional carbon transmitter does this with a power gain on the order of 20 dB. (The energy for supplying this gain comes from the direct current flowing through the carbon element.)

The *earpiece,* or *receiver,* converts electrical energy into acoustical energy. A passive device, the receiver performs this conversion with some loss.

In the 1950s, the *type 500 telephone set* was introduced. Although this set has been enhanced over the years with sleeker housings, multiline capability, and pushbutton dials, its transmission characteristics have remained substantially unchanged. During these decades, the evolution of the telephone network has been influenced by the performance of the 500 set. Today, the transmission characteristics of new telephone set designs must approximate those of a 500 set in order for the new designs to be compatible with the network. The Electronics Industries Association publishes EIA-470, which sets performance and compatibility standards for telephone sets.[5]

2-2-3 Local Loops

Local loops are typically constructed of unamplified copper-wire pairs. The loop loss ranges from near 0 dB (for a subscriber very near the EO) to about 8 dB for subscribers at several miles distance. Finer wire gauges are used for subscribers near the EO and coarser gauges for distant subscribers. In general, loops are designed to minimize the amount of copper used while still providing less than 8 dB of loss. For very long loops, amplification is provided by EO-installed or field-installed voice-frequency repeaters. A combination of fine-gauge cable and repeaters can be economically used even for moderate length loops. Still another variation on local loop design is use of *station* or *subscriber carrier.* These multiplex systems serve multiple subscribers on each wire pair. Whatever the implementation, most subscriber loops have a loss between 1 and 8 dB.

2-2-4 Local Calls

Based on our model, two telephones connected together for a local call will experience a typical electrical loss between them of 2 to 16 dB. The two parties will still hear each other with adequate level (recall that each telephone has an acoustical-to-electrical gain of about 20 dB). Note that it is possible for a call to sound *too loud.* (In the absence of noise or other impairments, connections with very little loss sound subjectively worse than calls with moderate loss.)[13]

2-2-5 Zero Transmission Level Points

Over the range of talker levels, the output dynamic range of a 500 set has a high-end level around 0 dBm. This maximum power level will exist at the following points:

1. subscriber end of a local loop
2. EO end of a local loop (when connected via the EO to a zero-length loop)
3. EO end of a tandem-connecting trunk (also when connected via the EO to a zero-length loop)

Therefore, in our transmission plan we select a 0 TLP for these points. That is, both ends of a subscriber loop operate at 0 TLP. Also, a local office operates at 0 TLP.

2-2-6 Toll Calls and Echo

Due to the greater distances involved, intertandem and tandem-connecting trunks are usually implemented on carrier facilities. Since carrier systems have VF gain available, these trunks can operate at any convenient net loss. Ignoring other constraints, we would probably assign 0-dB loss to all trunks. This would allow toll calls to sound like local calls (as far as signal level goes). However, there is the important constraint of echo. As explained in more detail in Chapter 5, echo occurs in the network due to impedance mismatches. The greatest mismatch occurs in the EO at the boundary between the TCT and the local loop. The signal reflection from this point will be heard by the caller as *talker echo* (Fig. 2-12). The echo's delay is a function of the electrical length of the connection. Echo can be quite objectionable to a telephone user. If the echo has sufficient delay and amplitude, a talker may actually have trouble speaking. Reducing either the amplitude or the delay makes the echo less objectionable.

Echo Reduction

Improved impedance match.   One way to reduce the echo's amplitude is to improve the impedance match. However, the impedances of subscriber loops

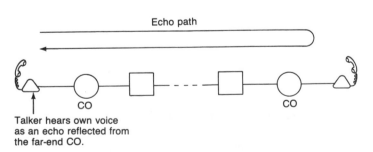

Echo path

Talker hears own voice
as an echo reflected from
the far-end CO.

FIGURE 2-12 Talker echo.

vary greatly due to the loops' infinite variety of construction. Historically, the best match has been a compromise that still results in considerable echo.

Added loss.   Another way to attenuate the echo is to add net loss to the toll transmission circuit. Trunks are designed to have equal loss in each direction. For example, a 2-dB loss added to a circuit would attenuate the desired signal by 2 dB, but would attenuate the *round-trip* signal (echo) by 4 dB.

Since longer circuits have longer delays, more loss should be added to longer circuits to maintain uniform echo performance. There is a limit, however, on added loss. At a distance of around 1850 miles (for terrestrial circuits), the added attenuation of the desired signal becomes too great. Circuits longer than 1850 miles are treated with echo suppressors or echo cancelers.[13] No additional loss is added to these long circuits. Echo suppressors and cancelers serve to eliminate the echo altogether. Their operation is covered in Chapter 5.

Loss Plans

There are two plans for allocating the loss necessary to control echo. The older *via net loss* plan was developed for a network that used all-*analog* transmission and switching. The newer *fixed-loss* plan applies to a network using all-*digital* transmission and switching. The network currently is a combination of the two plans.

Via net loss plan.   The Via Net Loss (VNL) plan was in part based on subjective studies that showed the relationship among echo delay, echo amplitude, and performance. It was found that for best performance, built-up toll circuits should have a fixed loss of about 5 dB plus an additional loss that increased with distance.[13] This additional loss is called the *via net loss*. The 5 dB is allocated 2.5 dB to each tandem-connecting trunk. The VNL is allocated on a per-mile basis to all trunks, including TCTs. Thus, under the VNL plan the following losses are assigned to each trunk type:

Tandem-connecting trunk: 2.5 dB + VNL
Intertandem trunk: VNL
Intertandem trunk with
    echo suppressor or canceler: 0 dB

Even though the tandem-connecting trunk loss objective is 2.5 dB + VNL, short nonrepeatered TCTs are allowed a loss of 2 to 4 dB.

Figure 2-13a shows a typical toll call and the losses that would be assigned to the trunks. Since the VNL value assigned to each trunk is only 1 dB or so, the total trunk loss of this typical connection may be about 7 dB.

Fixed-loss plan.   In a digital network, it is not desirable to add small amounts of loss to each trunk. The fractional-dB digital attenuators required would cause slight distortion and this distortion would be cumulative over a number of trunks in tandem. Subjective studies made since the adaptation of the

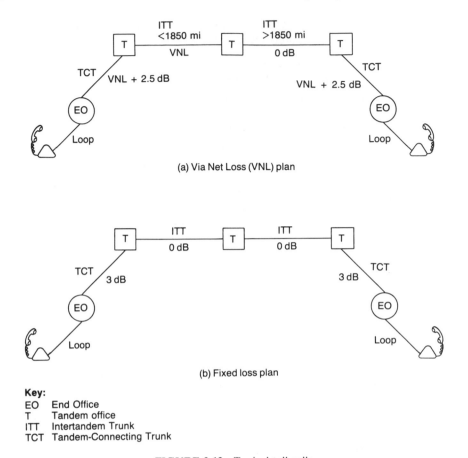

(a) Via Net Loss (VNL) plan

(b) Fixed loss plan

Key:
EO   End Office
T   Tandem office
ITT   Intertandem Trunk
TCT   Tandem-Connecting Trunk

FIGURE 2-13   Typical toll calls.

VNL plan show that it is possible to attain adequate performance by allocating *all*
the echo-controlling loss to the TCTs. The ITTs would operate at 0 dB loss. Echo
suppressors and cancelers would be applied as for an analog network. The loss
allocated to each TCT is 3 dB. Thus a toll call in a digital network has a fixed 6-dB
loss from EO to EO (Fig. 2-13b).

2-2-7 Tandem Office Transmission Level Points

Analog tandem TLP.   In our network, analog tandem offices are assigned a
−2 TLP. The model in Fig. 2-14a will help justify this assignment. Recall that EOs
are assigned a 0 TLP. Therefore, test tones leave end offices at 0 dBm. Trunks
connecting EOs and analog tandem offices are designed with a loss objective of
2.5 dB + VNL or 2 dB minimum. Thus test tones (and voice signals) arrive at
analog tandem offices at least 2 dB below the level at which they originate from

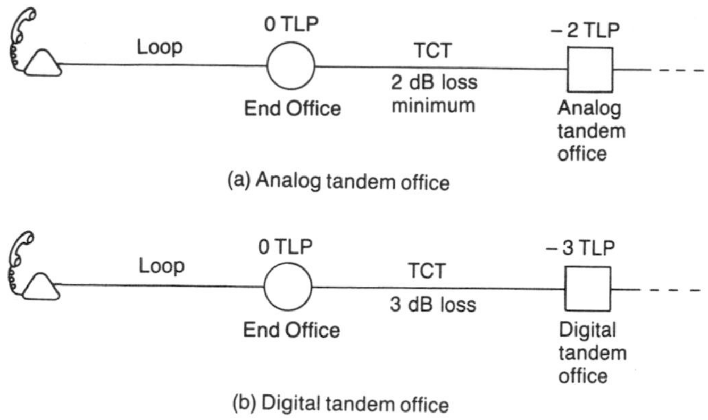

FIGURE 2-14   Transmission level points of tandem offices.

the EO. Tandem and end offices could operate at the same TLP. If this were the case, there would be an extra 2 dB of headroom at the tandem office. Extra headroom is fine, but the Signal-to-Noise (S/N) ratio at the tandem office would deteriorate as low level signals slip 2 dB into the noise. To maintain the S/N ratio in our developing transmission plan, analog tandem offices are operated at a −2 TLP.

Digital tandem TLP.   By the same reasoning, digital tandem offices are assigned a −3 TLP (Fig. 2-14b).

2-2-8 Test Pads and Measured Loss

Test Pads

A *test pad* is inserted between the test instruments and the office's test port (Fig. 2-15). The loss of the test pad in decibels is equal to the complement of the office TLP. To send a 0-dBm0 tone, the test oscillator is set at 0 dBm. The test pad attenuates the tone so that it is applied to the switch at the level of the office's TLP. To send −10 dBm0, just drop the oscillator to −10 dBm. Notice that you do not need to mentally juggle the office TLP into the considerations when setting the oscillator level. You can pretend that all offices are 0 TLP—the test pads will compensate for the cases of −2 and −3 TLP offices. Do not neglect to notice that there is also a test pad on the level meter.

Inserted Connection Loss

The trunk *loss* we have been discussing is more formally called *Inserted Connection Loss* (ICL). If you start with a built-up connection and add one more trunk in tandem, the additional loss added to the connection is equal to the added trunk's ICL.

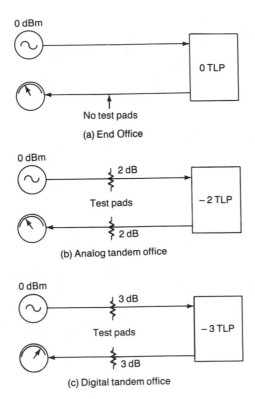

FIGURE 2-15 Test pads.

Inserted connection loss is used in network planning and design. It is not necessarily a loss that is measured directly.

Expected Measured Loss

Recall that trunk loss is measured through test pads at each end of the circuit under test. *Expected Measured Loss* (EML) is the trunk loss expected when the test pads are included (Fig. 2-16). Denoting the two test pad losses as TP1 and TP2, we obtain

$$EML = TP1 + ICL + TP2 \qquad (2\text{-}8)$$

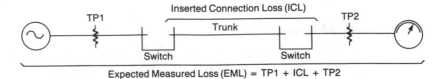

FIGURE 2-16 Expected measured loss.

Actual Measured Loss

Actual Measured Loss (AML) is the trunk loss actually measured and recorded in maintenance documents. The AML should approximately equal the EML. Typical maintenance practices dictate that corrective action be taken if the AML deviates from the EML by more than a decibel or so (see Table 2-10).

Note that *inserted* connection loss is not measured using the *insertion* loss definition of loss. Rather, ICL is measured using the *transducer* loss definition. The same applies for EML and AML.

2-2-9 Trunk Design

In this section we show example trunk designs to illustrate the concepts of TLP, ICL, EML, pads, and so on, just introduced.

Example 2-8

Figure 2-17a shows a tandem-connecting trunk implemented on a cable pair. The cable loss is 6 dB and the TCT connects to an analog tandem. Given an ICL of 3 dB, how is the trunk designed, and what is the EML?

The ICL is obtained by providing a repeater with a gain of 3 dB. Note that ICL = cable loss − repeater gain. From Eq. (2-8), the expected measured loss is

$$TP1 + ICL + TP2 = 0 \text{ dB} + 3 \text{ dB} + 2 \text{ dB}$$

$$= 5 \text{ dB}$$

It might be instructional to follow the 0-dBm test tone in the example from west to east as it is first amplified by the repeater, attenuated by the cable, then further attenuated by the test pad to arrive at the east level meter at −5 dBm. Now follow a tone from east to west. The tone's absolute levels along the way may not correspond to those of the west-to-east direction, but the tone still arrives at the west level meter at −5 dBm. Thus the EML is symmetrical (equal in both directions of transmission). Note also that the AML is read directly from the level meter by complementing the absolute reading. That is, an absolute level reading of −5 dBm corresponds to an AML of 5 dB.

Example 2-9

This example (Fig. 2-17b) has the same TCT as in Example 2-8, but now the trunk is implemented on carrier. Since the offices are two-wire and carrier is inherently four-wire, there are hybrids at each end of the trunk. (We have assumed a through loss of 3.5 dB for the hybrids in this series of examples.) The carrier has fixed transmit and receive TLPs of −16 dBm and +7 dBm. Design the trunk.

The choice of pad values sets the ICL at 3 dB. We will follow the rule that test tones are applied to carrier input (transmit) ports at the port's TLP (−16 dBm here). The ICL will be set by the pad at the *receive* end. The purpose of this rule is to maintain the signal-to-noise ratio in the carrier by operating at as high a level as possible.

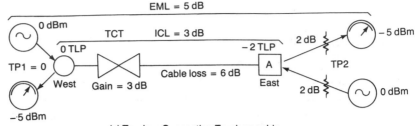

(a) Tandem-Connecting Trunk on cable

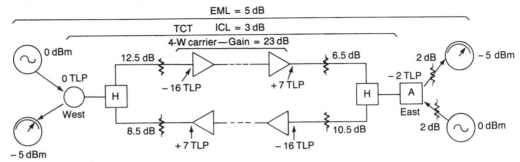

(b) Tandem-Connecting Trunk on four-wire carrier

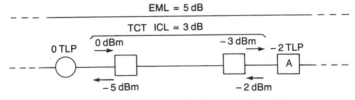

(c) Tandem-Connecting Trunk on two-wire carrier

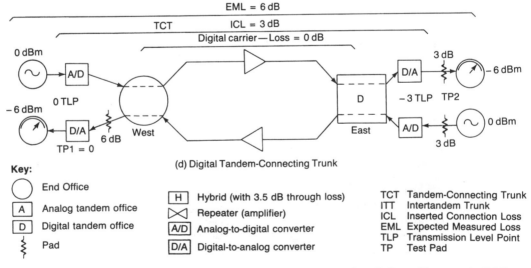

(d) Digital Tandem-Connecting Trunk

Key:

◯	End Office	H	Hybrid (with 3.5 dB through loss)
A	Analog tandem office	⋈	Repeater (amplifier)
D	Digital tandem office	A/D	Analog-to-digital converter
� Pad	Pad	D/A	Digital-to-analog converter

TCT Tandem-Connecting Trunk
ITT Intertandem Trunk
ICL Inserted Connection Loss
EML Expected Measured Loss
TLP Transmission Level Point
TP Test Pad

FIGURE 2-17 Example trunk designs. (Part (d) adapted from Bellcore *Notes on the BOC Intra-LATA Networks—1986.*[13])

Let's examine the west-to-east direction first. In order for the 0-dBm test tone to enter the carrier at −16 dBm, we must assign a pad loss of 12.5 dB:

$$0 \text{ dBm} - 3.5 \text{ dB (hybrid)} - 12.5 \text{ dB (pad)} = -16 \text{ dBm}$$

This tone will emerge from the east carrier channel at +7 dBm. After passing through another pad and the hybrid, the tone level should be −3 dBm (since the ICL = 3 dB). This requires a receive pad loss of 6.5 dB.

The east-to-west direction is designed in a similar manner. The east 0-dBm test tone is attenuated 2 dB in the test pad, 3.5 dB in the hybrid, and 10.5 dB in the transmit pad to enter the carrier at −16 dBm. The tone emerges in the west at +7 dBm. The 8.5-dB pad and west hybrid drop this to −5 dBm, as required.

In Fig. 2-17b we have shown the hybrids, pads, and carrier channel units as separate pieces of equipment. Transmission equipment used to be provided this way, resulting in much wiring between separate units. Today, the hybrid and pads would probably be combined into the channel units, which would now become *two-wire* channel units (Fig. 2-17c). The internal pads are adjusted to provide the test tone levels shown. Note that the standard +7 and −16 access points may be lost in this arrangement.

For another illustration of trunk design, Fig. 2-17d shows a *digital* tandem-connecting trunk. Digital trunks connect digital offices via digital carrier. The carrier may be carried on wire pairs, coax, fiber optic cable, or digital microwave radio.

The fixed loss plan calls for the ICL of digital TCTs to be 3 dB. In this example,

$$EML = TP1 + ICL + TP2 = 0 + 3 \text{ dB} + 3 \text{ dB}$$
$$= 6 \text{ dB}$$

Let's follow the test tone from west to east. The 0-dBm test tone encodes in the west A/D converter at a level equal to the digital milliwatt. There is no attenuation in the digital carrier. The tone decodes at the D/A converter in the east test port. The decode level is −3 dBm, which corresponds to the −3 TLP of this office. The test tone next passes through a 3-dB test pad to arrive at the level meter at −6 dBm, corresponding to an EML of 6 dB.

In the east-to-west direction, the test tone is first attenuated 3 dB by the transmit test pad. The resulting −3-dBm signal is encoded at a level equal to the DMW. At the west EO, the stored program control detects that this connection is to a digital TCT and switches in a 6-dB digital pad. Since the west D/A converter is aligned to decode the DMW at 0 dBm, it decodes the digitally attenuated test tone at −6 dBm. The level meter displays −6 dBm, corresponding to an EML of 6 dB.

Note that the 6-dB pad at the west EO would be switched in for any call that connects a local subscriber to a digital TCT. That is, the 6-dB pad is a normal transmission pad used on real calls; it is not a test pad. (There are no test pads at the west EO since its TLP = 0.) The 6-dB end-to-end loss for toll calls in the fixed loss plan is set by this lone pad in the receive end of the built-up connection.

The examples given above are just a sampling of trunk design variations. There may be infinite trunk designs as analog, digital, end-office, tandem, two-wire, and four-wire equipment is combined. Bellcore *Notes on the BOC Intra-LATA Networks—1986*[13] contains more design examples.

2-2-10 Level Limits

Excessive signal levels can crosstalk to other circuits, overload analog carrier systems, cause distortion, or simply sound too loud. Table 2-7 lists signal level limits for subscriber equipment and Table 2-8 lists additional miscellaneous limits. Note that short-duration signals, as used for signaling and testing, are allowed higher levels than longer-duration signals such as voice.

2-2-11 Loss Objectives and Performance

Subscriber Loops

Objectives. Table 2-9 lists the objectives for subscriber loop loss. Notice that net gain is not allowed on subscriber loops (i.e., the minimum loss is 0 dB). Also note that for economic reasons, longer loops are allowed more loss.

Performance. Subscriber loop loss performance data are available from Bell's 1973 Loop Survey.[23] This survey found the average 1000-Hz loop loss to be 3.7 dB. It also found that 95% of main stations are served by loops with 1000-Hz loss of less than 7.5 dB. (Later surveys, such as the 1980 Noise Survey, show that these early results are still valid.) This reported loop performance falls well within the recommendations of Table 2-9.

Trunk Objectives

The loss objectives for trunks are functions of the loss plans as described earlier. In addition to loss objectives, there are also trunk loss *accuracy* limits, as listed in Table 2-10.

Switching and Subscriber Equipment Objectives

Tables 2-11 and 2-12 list the through-loss objectives for switching and subscriber equipment. Not listed are the FCC's complex requirements for the through loss of terminal equipment. Basically, the FCC attempts to prevent *gain* on a through-switch connection. You are referred to FCC Part 68.308 for details.

Total Network Performance

Loss performance for the total network is available from Bell's 1982/83 End Office Connection Study (EOCS)[22] and Bell's 1980 Noise Survey.[21] The EOCS found the mean 1004-Hz end office-to-end office loss to be 6.8 dB. About 99% of such calls had a 1004-Hz loss lower than 12 dB.

When the EOCS loss distribution and the subscriber loop loss distribution (from the 1980 Noise Survey) were combined, the EOCS authors found that about 95% of customer premises-to-customer premises calls had less than 23 dB of loss. Only about 5% of such calls had a loss lower than 10 dB.

2-3 FREQUENCY RESPONSE

Frequency response is the gain or loss of a transmission path expressed as a function of frequency. (Frequency response is also called *amplitude distortion.*) In audio work, frequency response is usually expressed (in dB) as the *gain* at the frequency of interest relative to the gain at 1000 Hz. It is also possible to express frequency response in terms of *loss,* and many telephone standards do this. As a further variation, loss response can be plotted with values of increasing loss either ascending or descending the vertical axis. Some telephony response curves may appear upside down to you if your audio background includes viewing *gain* response plots in the consumer audio literature. In telephony, frequency response is usually measured at 0 dBm0 (test tone level).

2-3-1 Bandwidth and Cutoff Frequencies

We define *bandwidth* as the frequency band between lower and upper *cutoff frequencies* where the loss at the cutoff frequencies is *10 dB* greater than the loss at 1004 Hz. Since *3-dB* cutoff points are also common in audio work, we will remind you of our use of 10-dB points each time we discuss bandwidth.

2-3-2 Slope

Slope (or *gain slope*) is a visually descriptive term that describes a frequency response characteristic. By definition, *404-Hz slope* is the loss at 404 Hz relative to the loss at 1004 Hz. The *2804-Hz slope* is the loss at 2804 Hz relative to the loss at 1004 Hz. The three frequencies 404, 1004, and 2804 Hz are often available as preset frequencies on test oscillators. Slope measurements are typically made at −16 dBm0.

2-3-3 Frequencies to Avoid

Submultiples of 8000 Hz

We have already mentioned that a test frequency of exactly 1000 Hz should be avoided when measuring levels on digital systems. When measuring frequency response, other exact submultiples of 8000 Hz should also be avoided for the same reason. This is why test oscillators with preset frequencies usually offset these frequencies by 4 Hz to produce 404 Hz, 1004 Hz, 2004 Hz, and so on.

Near 2600 Hz

It is also best to avoid testing at or near 2600 Hz when checking circuits working through the telephone network. Real circuits often pass through Single-Frequency (SF) signaling sets. These sets use a single frequency (usually 2600 Hz) to signal on-hook and off-hook (and sometimes dial pulses) from one end of a circuit to the other. Absence of the 2600-Hz tone indicates off-hook. Tone present indicates on-hook. The sets cannot distinguish between a valid signaling tone and a test tone. If you tune through 2600 Hz while measuring frequency response, the far-end signaling set will think you have gone on-hook (hung up) and will drop the circuit. Sweep oscillators used in telephony often have an *SF skip* switch, used to skip 2600 Hz while sweeping.

2-3-4 Frequency Response Objectives and Performance

Subscriber Loops

Objectives.   ANSI/IEEE Std. 820-1984, *IEEE Standard Telephone Loop Performance Characteristics*[9] specifies the frequency response for subscriber loops. A summary of this specification is provided in Table 2-13 at the end of the chapter. The IEEE *recommends* a minimum bandwidth (between 10-dB points) of 3000 Hz (i.e., 200 to 3200 Hz). The IEEE standard further states that a 2200-Hz bandwidth (300 to 2500 Hz) is *acceptable*.

Performance.   Data on subscriber loop frequency response performance are available from the Bell System's 1973 Loop Survey.[23] Table 2-2 is derived from that survey. The mean frequency response performance fits well within the objectives of Table 2-13.

Carrier System and Switching Equipment Objectives

Frequency response objectives for carrier systems and switching equipment are listed in Tables 2-14 and 2-15. Note that the standards for digital systems call for purposely limited high-frequency response. This equipment uses Pulse Code Modulation (PCM), and the frequency response of the channels must be coordi-

TABLE 2-2 FREQUENCY RESPONSE
PERFORMANCE: SUBSCRIBER LOOPS

Adapted from the Bell System 1973 Loop Survey[a]	
Frequency (Hz)	Mean loss[b] (dB)
200	−0.8
300	−0.9
500	−0.7
1500	0.9
2000	1.7
2500	2.7
3000	3.8
3200	4.6

[a] Reprinted with permission from AT&T.
Copyright 1977 AT&T.

[b] Relative to mean loss at 1000 Hz.

nated with the sampling rate of the PCM coders. Since this rate is 8000 samples per second, the highest voice frequency that will be correctly encoded is 4000 Hz. (We discuss PCM systems in more depth in Chapter 7.) Before encoding at the transmitting end, voice signals must be bandlimited to below 4000 Hz by a low-pass *presampling filter*. Practical filter designs begin rolling off around 3400 Hz in order to provide sufficient attenuation above 4000 Hz. Thus the high end of a digital system's response is around 3400 Hz.

At the receiving end, a PCM decoder converts the digitized voice back to an analog signal. Due to the nature of the sampling process, the recovered analog signal contains unwanted components that start at 4000 Hz in frequency and extend high above the voice band. Another low-pass filter, the postsampling filter, must be used to eliminate these components. The postsampling filter can have characteristics similar to those of the presampling filter. In fact, it is generally assumed that at each frequency, the end-to-end frequency response of a PCM coder/decoder pair is divided in amplitude equally between the two ends. This can be seen in Table 2-14. A major exception to this equal division occurs at 60 Hz.

At the low end of the band, the PCM process imposes no frequency response limitation. In theory, response to dc is possible. For practical circuit reasons, however, most voice circuits are ac-coupled via transformers or capacitors. Low-end performance need not be extended much below 200 or 300 Hz.

Many telephone circuits have a large 60-Hz component that is coupled from nearby ac power lines. This 60-Hz noise is superimposed on the voice signal. If allowed into the PCM coder (A/D converter), the 60-Hz component would offset the voice signal within the encoding range. This offset would occur at a 60-Hz rate and could produce audible components from the otherwise inaudible 60-Hz noise.

To alleviate this problem, the loss of digital systems is required to be around 20 dB at 60 Hz. Note that the loss at 60 Hz must be in the circuit at the *transmit end* and *before* the coder.

Subscriber Equipment Objectives

Frequency response objectives for key telephone systems, Private Branch Exchanges (PBXs), and other terminal equipment are listed in Table 2-16. Only the PBX standard specifies a flat (nearly) frequency response. The other standards simply seek to protect two frequencies that have special uses in the network.

As mentioned previously, tones at 2600 Hz are used for single-frequency signaling. Normally, it takes fairly pure tones to activate SF sets; the sets are relatively immune to the wide energy spectrum of voice. However, if the frequency response of terminal equipment should be skewed to favor 2600 Hz, the probability of SF *talk-off* increases. (Talk-off occurs when an SF set detects voice as a tone and drops the connection.) The subscriber equipment standards require that the band 2450 to 2750 Hz not be accentuated more than 1 dB relative to the rest of the voice band.

The other frequency protected by the standards is 4000 Hz, which is used as a *pilot tone* by some analog carrier systems. (Pilot tones serve to regulate repeater gain.)

TABLE 2-3 FREQUENCY RESPONSE PERFORMANCE: TOTAL NETWORK

Adapted from the Bell System 1982/83 End Office Connection Study[a]		
	Loss[b] (dB)	
Frequency (Hz)	Short connections	Medium distance and long connections
204	5.1	5.8
304	1.8	2.1
404	1.1	1.4
604	0.4	0.5
1004	0.0	0.0
2004	0.1	0.1
2804	1.7	0.8
3004	2.3	1.1
3204	4.1	2.2
3404	7.4	7.5
3504	16.5	19.2

[a] Reprinted with permission from AT&T. Copyright 1984 AT&T.

[b] Mean loss relative to mean loss at 1004 Hz. Measured end office to end office.

Total Network Performance

Frequency response performance for the total network (end office to end office) is available from Bell's 1982/83 End Office Connection Study (EOCS).[22] Some of the data from that study are presented in Table 2-3. The authors of the EOCS point out that the network's frequency response approximates the response of the widely used PCM carrier systems (compare Tables 2-3 and 2-14). As the network continues its digital evolution, this approximation will tighten.

2-4 REFERENCE TABLES

TABLE 2-4   OVERLOAD OBJECTIVES

From Bellcore *Notes on the BOC Intra-LATA Networks—1986*[13]	
Analog trunks	+5 dBm0
Digital switches	+3 dBm0

From AT&T PUB 43801, *Digital Channel Bank Requirements and Objectives* (1982)[12]
+3 dBm0

From REA 522 (Digital Central Office)[14]
+3 dBm0

From EIA-464 (PBX)[4]	
Input level (dBm)	Maximum compression[a] (dB)
0	0.1
+4	0.1
+7	0.4

[a] Must be met by 95% of cross-office connections (station to station, station to trunk, and trunk to trunk). Compression is a reduction in gain relative to the gain at an input level of −9 dBm; measured at 1000 Hz. This specification, which indicates an overload point near +7 dBm, is under review for its applicability to digital PBXs.

TABLE 2-5 LEVEL TRACKING OBJECTIVES

From AT&T PUB 43801, *Digital Channel Bank Requirements and Objectives* (1982)[12]			
	Maximum level deviation[b] (dB)		
	End to end		One end only
Input level[a] (dBm0)	Any channel	Averaged over ≥20 banks	Any channel
+3 to −37 −37 to −50	±0.5 ±1	±0.25 ±0.5	±0.25 ±0.5

From REA 522 (Digital Central Office)[14]	
Input level[a] (dBm0)	Maximum level deviation[b] (dB)
+3 to −37 −37 to −50	±0.5 ±1

[a] At 1004 Hz.

[b] Deviation from gain at 0 dBm0.

TABLE 2-6 LEVEL STABILITY OBJECTIVES

From AT&T PUB 43801, *Digital Channel Bank Requirements and Objectives* (1982)[12]	
Condition	Level stability[a]
After 20 years	±0.5 dB

From REA 522 (Digital Central Office)[14]	
Condition	Level stability[b]
In both directions of transmission	±0.5 dB

[a] Measured between 1004 and 1020 Hz at 0 dBm0. Must be met by 95% of circuits.

[b] Measured at 1004 Hz, at 0 dBm0, and at 20 to 30°C.

TABLE 2-7 LEVEL LIMITS: SUBSCRIBER EQUIPMENT (VOICE BAND METALLIC)

Composite from EIA-470 (Telephone Instruments),[5] EIA-478 (Key Telephone Systems),[6] EIA-464 (PBXs),[4] and FCC Part 68 (Terminal Equipment)[1]		
Condition	Interface	Maximum signal level[a] (dBm)
Nonlive voice, nondata, nonsignaling[b]	CO trunk/line	−9
	Two-wire tie trunk	−15
	Four-wire lossless tie trunk	−15
	Four-wire CTS[c] tie trunk	−19
	Off-premises station	−13
Signaling[d]	CO trunk/line	0
	Two-wire tie trunk	−4
	Four-wire lossless tie trunk	−4
	Four-wire CTS[c] tie trunk	−8
Data	Fixed-loss loop	−4
	Programmed	−12 to 0
	Nonadjustable	−9

[a] Into 600 Ω, in the band 200 to 4000 Hz, with dc from loop simulator (see Chapter 1), averaged over 3 seconds.

[b] Such as music-on-hold and recorded announcements.

[c] A four-wire *CTS* (Conventional Terminating Set) trunk circuit contains a built-in hybrid with an assumed through-hybrid loss of 4 db.

[d] Such as DTMF.

TABLE 2-8 LEVEL LIMITS: MISCELLANEOUS (VOICE BAND METALLIC)

From Bellcore *Notes on the BOC Intra-LATA Networks—1986*[13]	
Condition	Maximum signal level (dBm0)
Average long-term input power per carrier channel	−16
Average over 3 seconds on an actual call	−13
Short-term test tones	0
Long-term test tones	−10
Test tones from new automatic test systems	−16

TABLE 2-9 LOSS OBJECTIVES: SUBSCRIBER LOOPS

From Bellcore *Notes on the BOC Intra-LATA Networks—1986*[13]
8.5 dB maximum loss at 1000 Hz

Adapted from ANSI/IEEE 820-1984, *IEEE Standard Telephone Loop Performance Characteristics*[9]		
	Loss[a] (dB)	
Loop length (1000 ft)	Recommended	Acceptable
0	0–5.5	0–7
6	0.3–6.3	0–8.2
12	0.7–7.2	0–9.3
≥18	1–8	0–10.5

[a] 1004-Hz transducer loss (600 Ω at subscriber and 900 Ω at CO). The *preferred* loop loss for any length is 3.0 to 4.0 dB.

TABLE 2-10 LOSS ACCURACY OBJECTIVES: TRUNKS

Adapted from Bellcore *Notes on the BOC Intra-LATA Networks—1986*[13]		
	Maximum loss deviation[a] (dB)	
Trunk makeup	Preservice	Immediate action[b]
Nonrepeatered interbuilding cable	±1.0	±3.7
All others	±0.5	±3.7

[a] Deviation of actual measured loss from expected measured loss at 1004 Hz.

[b] Circuits not meeting an *immediate action* limit must be taken out of service or repaired immediately.

TABLE 2-11 LOSS OBJECTIVES:
SWITCHING EQUIPMENT

From REA 522 (Digital Central Office)[14]	
Condition	Loss[a]
Trunk to trunk or trunk to line Line to line	0–0.5 dB 0–2 dB

From REA 524 (Analog Central Office)[15]	
Condition	Loss[b]
Cross-office	0–0.5 dB

[a] Measured at 1004 Hz, at 0 dBm input, in both directions of transmission, at 20 to 30°C, and at 600 or 900 Ω, as required to match impedances.

[b] Measured at 1000 Hz, at 0 dBm input, at 900 Ω, at 20 mA dc, and at 70 mA dc. (REA 524 also specifies insertion loss at other frequencies; see Table 2-15.)

TABLE 2-12 LOSS OBJECTIVES: SUBSCRIBER EQUIPMENT

From EIA-464 (PBXs)[4]		
	Loss[a] (dB)	
Connections	Desired	Required
On premises-to-on premises stations	4.5–5.5	0–6
On premises-to-off premises and off premises-to-off premises stations	0–0.5	0–6
Trunk to station without pads[b]	—	0–0.6
Trunk to station with pads	—	1.8–2.8
Trunk to trunk, four-wire without pads	—	0
Trunk to trunk, four wire equipped with pads, but pads switched out	—	0.5
Four-wire trunk with pads to two-wire trunk	—	1.8–2.7

[a] Insertion loss at 1004 Hz averaged over all connections of the same type. Test instruments set to match associated PBX port impedance.

[b] Refers to 2-dB *switched pads* that may be inserted into some connections.

TABLE 2-13 FREQUENCY RESPONSE OBJECTIVES: SUBSCRIBER LOOPS

Adapted from ANSI/IEEE 820-1984, *IEEE Standard Telephone Loop Performance Characteristics*[9]		
Condition	Frequency (Hz)	Recommended loss[a] (dB)
Each direction; 600-Ω equipment at subscriber end, 900-Ω equipment at CO end	200	−3.5 to 10
	300–500	−3 to 3.5
	800–1500	−1.25 to 3.5
	3200	−1.25 to 10
	3400	−1.25 to 22.5
	4000	≥−1.25

[a] *Transducer* loss relative to loss at 1004 Hz. This table is adapted from the curves in IEEE 820, which also show looser *acceptable* limits.

TABLE 2-14 FREQUENCY RESPONSE OBJECTIVES: CARRIER SYSTEMS

Adapted from *Digital Channel Bank Requirements and Objectives*[12] (PUB 43801)[a]				
Circuit type	Frequency (Hz)	Loss[b] (dB)		
		Transmit end	Receive end	End to end
Two-wire	≤60	≥20	≥0	≥20
	200	0 to 3	0 to 2	0 to 5
	300–3000	−0.25 to 0.5	−0.25 to 0.5	−0.5 to 1
	3200	−0.25 to 0.75	−0.25 to 0.75	−0.5 to 1.5
	3400	−0.25 to 1.5	−0.25 to 1.5	−0.5 to 3
	4000	≥14	≥14	≥28
	≥4600	≥32	≥28	≥60
Four-wire	≤60	≥14	≥0	≥14
	200	−0.15 to 2	−0.15 to 1	−0.3 to 3
	300–3000	±0.15	±0.15	±0.3
	3200	−0.15 to 0.75	−0.15 to 0.75	−0.3 to 1.5
	3400	−0.15 to 1.5	−0.15 to 1.5	−0.3 to 3
	4000	≥14	≥14	≥28
	≥4600	≥32	≥28	≥60

[a] Reprinted with permission of AT&T. Copyright AT&T 1982; all rights reserved.

[b] Relative to loss at 1000 Hz. Measured at 0 dBm0. This table is derived from the frequency response curves given in PUB 43801.

TABLE 2-15 FREQUENCY RESPONSE OBJECTIVES: SWITCHING
EQUIPMENT

From REA 522 (Digital Central Office)[14]		
Condition[a]	Frequency (Hz)	Loss[b] (dB)
Line to line	60	≥20[c]
	300	−1 to 3
	600–2400	±1
	3200	−1 to 3
Two-wire trunk to two-wire trunk	60	≥20[c]
	200	0 to 5
	300–3000	−0.5 to 1
	3300	≤1.5
	3400	0 to 3
Four-wire trunk to four-wire trunk	60	≥16[c]
	200	0 to 3
	300–3000	±0.3
	3300	≤1.5
	3400	0 to 3

From REA 524 (Analog Central Office)[15]		
Condition	Frequency (Hz)	Loss[d] (dB)
Capacitor-coupled trunk circuits	300	0–1.0
	500	0–0.5
	3400	0–0.5
Transformer-coupled trunk circuits	300	0–1.5
	500	0–1.0
	3400	0–0.5

[a] Line-to-trunk frequency response should be a compromise between the responses for line to line and trunk to trunk.

[b] Relative to loss at 1004 Hz. Measured at 0 dBm0.

[c] The loss at 60 Hz should all be in the transmit end of the connection.

[d] Relative to loss at 1000 Hz as derived from the REA 524 insertion loss specification.

TABLE 2-16 FREQUENCY RESPONSE OBJECTIVES: SUBSCRIBER EQUIPMENT

From EIA-464 (PBXs)[4]			
Condition	Frequency (Hz)	Loss[a] (dB)	
		Required	Desired
Station to station	200	−0.1 to 1.6	−0.1 to 0.6
	300	−0.1 to 1.0	−0.1 to 0.4
	3000	−0.1 to 1.0	−0.1 to 0.4
	3400	−0.1 to 1.5	−0.1 to 0.6
Station to trunk or two-wire trunk to four-wire trunk	200	−0.1 to 1.0	−0.1 to 0.3
	300	−0.1 to 0.8	−0.1 to 0.2
	3000	−0.1 to 0.8	−0.1 to 0.2
	3400	−0.1 to 1.0	−0.1 to 0.3
Four-wire trunk to four-wire trunk	200	−0.1 to 0.6	−0.1 to 0.3
	300	−0.1 to 0.4	−0.1 to 0.2
	3000	−0.1 to 0.2	−0.1 to 0.2
	3400	−0.1 to 0.3	−0.1 to 0.3

On a connection from any port, through the PBX, to a CO trunk, the loss at any frequency in the band 600 to 4000 Hz shall not exceed the loss at any frequency in the band 3995 to 4005 Hz by more than 3 dB.

From FCC Part 68 (Terminal Equipment)[1]
On a connection through the terminal equipment, the loss at any frequency in the band 800 to 2450 Hz shall not exceed the loss at any frequency in the band 2450 to 2750 Hz by more than 1 dB.
On a connection through the terminal equipment, the loss at any frequency in the band 600 to 4000 Hz shall not exceed the loss at any frequency in the band 3995 to 4005 Hz by more than 3 dB.

[a] Relative to loss at 1000 Hz. Must be met by 95% of connections. Under review for applicability to digital PBXs.

3 NOISE and CROSSTALK

Noise and crosstalk are two impairments that can degrade the performance of voice-frequency telephone circuits. When the circuit is used for voice transmission, these impairments can be objectionable to the talker and listener. When the circuit is used for voice-band data, these impairments can cause data errors (also objectionable to the end user).

In this chapter we divide noise into the broad categories of *steady noise* and *impulse noise*. Crosstalk is covered as a third major topic. These impairments can be categorized in several ways. Categorized by source, for example, noise can be attributed to various physical sources such as the random motion of electrons or the intermittent contact of a corroding wire splice. Noise can also be categorized by its amplitude and frequency distributions. (White noise has a frequency distribution quite different from single-frequency noise.)

When we consider the measurement of noise and crosstalk, we are again led to a matrix of categories. For example, noise can be measured

1. At a single frequency, or over a frequency band
2. In a balanced circuit, or with respect to ground
3. With or without a holding tone present

In this chapter we discuss noise measuring sets and show several of their applications.

The industry has published noise and crosstalk objectives for various elements of the network. These are summarized in the text and grouped in reference tables at the end of the chapter. Since objectives have not been set for some impairments, we can look at performance surveys for more data. It is also interesting to compare published objectives and measured performance.

3-1 STEADY NOISE

3-1-1 General Characteristics

Steady noise (also called *circuit message noise*) is usually heard as a hiss or a hum. Even as the network evolves toward digital transmission on optical fibers, most users recognize the characteristic rushing, hissing noise of a long-distance call as soon as they put the receiver to their ear. Hum, on the other hand, is usually a local phenomenon caused by interference from power lines. A subscriber on a long, rural loop might experience this problem. There are other sources of steady noise, but their common characteristic is that they both sound steady to the listener and appear steady when observed on a measuring instrument. This is in contrast to the nonsteady nature of clicks and pops which we call impulse noise and discuss later in this chapter.

Steady noise can be objectionable to listeners, especially during quiet pauses when no one is talking. (When speech is present it tends to mask the noise, making it less objectionable.) If high enough, steady noise can also cause errors in voice-band data. In general, steady noise is more a problem for listeners than for modems. (The converse is true for impulse noise, which can be more troublesome for modems than humans.)

The total steady noise on a telephone circuit is the sum of various steady noise sources. The nature of steady noise is that these sources are usually uncorrelated. Thus the individual noise sources are summed on a power basis to determine the total steady noise.

Sources of Steady Noise

Thermal noise.   Thermal noise is also called *Johnson* or *resistance noise*. It is caused by the random motion of electrons in conductors. It is white noise (equal energy at all frequencies) and increases with temperature.

Shot noise.   *Shot noise* is caused by the discrete nature of electron flow in active devices. It is white noise and is not a direct function of temperature.

Low-frequency noise.   Also called *1/f noise, low-frequency noise* has increasing amplitude at lower frequencies. It is caused by the fluctuations in conductivity of semiconductor material. In modern devices, 1/f noise is not a major concern.

Thermal, shot, and low-frequency noise sources exist to some extent in all circuits. In practice, these noise sources cause the most problems in circuits that process low-level signals. Examples include amplifiers operating at voice, carrier, and microwave frequencies. These noise sources usually sound like hiss.

Single-Frequency noise.   Single-Frequency (SF) noise is not a true noise source, but rather, a description of noise induced in a circuit from an interfering source. If the interfering source is a single frequency, the resulting noise will appear at the same frequency. If the interfering source also contains harmonics, so will the noise. A classic example is power-line hum, consisting of the harmonics of 60 Hz. In equipment where Voice-Frequency (VF) and digital signals are in close proximity, a digital signal may couple into a voice signal. If the digital signal contains a VF component, single-frequency noise will result. An example of such a digital signal is the 667-Hz signaling-frame clock present in D-type carrier channel banks. Inadequate circuit spacing could allow this signal to be heard on one or more channels as a 667- or 1333-Hz tone. Single-frequency noise can be coupled by electromagnetic fields or conducted via poorly isolated and filtered power supplies or talk batteries.

Impulse noise.   High-level impulse noise is treated later in this chapter in the section on impulse noise. Low-level and continuous impulses blend with other steady noise sources. Sources of impulse noise (at high or low levels) include arcing relay contacts, corroding connections, and bad wire splices. Impulses can be radiated to other circuits or conducted via shared circuits such as power supplies and talk batteries.

Quantizing noise.   Quantizing noise results from the digitization of speech for transmission over digital carrier. For example, in the Pulse Code Modulation (PCM) method of digitization, the incoming analog voice is sampled at intervals. Each sample amplitude is then quantized to one of 255 discrete amplitudes. The amplitude of each sample is sent to the far end encoded as an 8-bit word. At the far end, the signal is decoded into an analog signal. However, only 255 discrete amplitudes are available for decoding each sample received. The reconstructed analog signal will not exactly match the original signal, due to the error caused by quantizing the signal into discrete steps. This error is called the *quantizing noise*.

Quantizing noise is present only when a signal is present. Thus when quantizing noise is measured, a test signal must be applied to the circuit under test. At the far end, the test signal is removed and the residual noise level is measured. One way to measure quantizing noise is to perform a *C-notched noise* measurement. This measurement uses a fixed-level, sine wave, test signal called a *holding tone*. The holding tone is removed at the far end by using a sharp notch-reject filter. (C-notched noise is described in more detail later in this chapter.) When quantizing noise is measured using the C-notched technique, the result is expressed as a single absolute noise level.

Quantizing noise is the same impairment as quantizing distortion. The term *quantizing distortion,* however, implies a series of measurements made over a range of input signal levels. Each measurement is expressed as a signal-to-distortion ratio. Quantizing distortion and its measurement method are described in more detail in Chapter 7.

Aliasing noise. *Aliasing* is a phenomenon that occurs in sampled data systems. Aliasing turns an input signal of one frequency into an undesired output signal at another frequency. Since a signal must be present for aliasing to occur, it is usually considered to be a distortion impairment. We cover aliasing distortion in detail in Chapter 7.

Consider now a combined case of single frequency noise interference (described above) and aliasing distortion. An interfering noise source may couple into a circuit at a benign frequency, then via aliasing, appear as steady noise at some more annoying frequency. We have dubbed this phenomenon *aliasing noise.* An example is provided in Chapter 7.

Crosstalk noise. Crosstalk is a desired signal in the wrong place. The subject of crosstalk between selected individual circuits is covered as a major section later in this chapter. Here we list crosstalk noise as steady noise interference from many crosstalk sources and at low, unintelligible levels. It is a babbling effect that blends in with the other steady noise sources.

Analog-to-Digital converter noise. Analog-to-Digital (A/D) converter noise is not a noise source but rather, a phenomenon that can amplify an existing low-level noise. Imagine an A/D converter whose input is biased just at a decision point. Furthermore, let this be the decision between the step that encodes zero volts input and the step that encodes the first miniscule positive voltage. Now imagine the slightest amount of noise present at the input. The noise will cause the Least Significant Bit (LSB) of the output word to toggle between 0 and 1. At the far-end digital-to-analog converter, the 0-to-1 transitions of the LSB cause the converter to output a waveform with amplitude equal to one step. The step has a very small but finite voltage. (There are only 255 steps available, so they cannot be infinitely small.) If the output step is larger than the input noise, the noise has been amplified, and nonlinearly. This phenomenon establishes a *noise floor* for PCM systems. Steady noise performance better than the noise floor cannot be guaranteed.[16,17,26]

Steady Noise Reduction

Steady noise is minimized by careful design and construction of telephone network elements ranging from semiconductors to outside plant. Some examples follow.

Power-line interference is reduced by using highly balanced cable pairs and terminating equipment. Cable pairs must be carefully spliced to maintain good balance. (Chapter 6 discusses balanced circuits and their requirements.) Proper shielding, bonding, and grounding of outside plant also reduces power-line noise. Physical separation of telephone and power cables also helps.

Communications systems should be coordinated with noise reduction in mind. For example, digital and analog carrier systems may have to be segregated into separately shielded cables to avoid interference of the analog by the digital. Two other examples of coordination are the assignments of frequencies used in radio systems such as microwave relay and mobile telephone.

Noise reduction efforts are also important in inside plant design. In office layout it may be desirable to run signaling leads in cables separate from VF pairs. Of course, the CO talk battery must be well filtered, usually with inductor/capacitor banks distributed throughout the office.

In equipment design, Printed Circuit Board (PCB) layout is often critical to eliminating crosstalk and noise interference. Power supply bypassing (filter capacitors at the PCB and integrated circuit level) is also important, especially in hardware combining analog voice and digital circuits.

3-1-2 Frequency Characteristics

C-Message Weighting

The human ear is not equally sensitive to all voice frequencies. Nor do telephone sets have a flat electrical-to-acoustical frequency response. As a result, not all frequency components of noise are equally objectionable. In the early 1960s, Bell Labs updated its noise weighting curve to include the frequency response of the ear plus the receive response of a 500-type telephone set. This curve is called the *C-message weighting* characteristic (Fig. 3-1). It is used to shape the frequency spectrum of measured noise so that the measurement will correlate with the listener's subjective assessment. The curve has a 0-dB reference point at 1000 Hz. When C-message weighting is used, components both above and below 1000 Hz are attenuated to match the reduced high- and low-frequency response of the ear and telephone set.

3-kHz Flat Weighting

Even though the ear and telephone set may not be sensitive to low frequencies, strong low-frequency noise components may disrupt equipment operation. To check for this possibility, a *3-kHz flat weighting* curve is used (Fig. 3-2). This curve is made flat down to 60 Hz so that power-line interference can be assessed.

Measurement Units

dBrn.   The basic voice circuit noise measurement unit in telephony is the *dBrn*. The letters dBrn stand for decibels above *reference noise*. Reference noise is a power of -90 dBm (i.e., 0 dBrn = -90 dBm). Since reference noise is such a low level (1 picowatt or 10^{-12} watt), almost all noise measurements that are expressed in dBrn will be positive values.

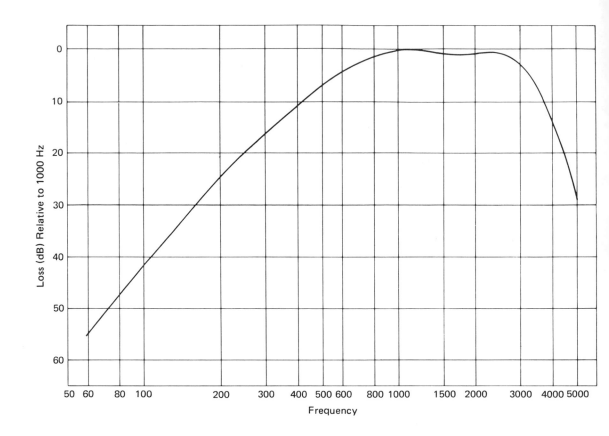

Tolerance (NOTE (1))
60 to 300 Hz ± 2 dB
300 to 1000 Hz ± 1 dB
1000 Hz 0
1000 to 3000 Hz ± 1 dB
3000 to 3500 Hz ± 2 dB
3500 to 5000 Hz ± 3 dB

Frequency	Design Loss (dB)
60	55.7
100	42.5
200	25.1
300	16.3
400	11.2
500	7.7
600	5.0
700	2.8
800	1.3
900	0.3
1000	0
1200	0.4
1300	0.7
1500	1.2
1800	1.3
2000	1.1
2500	1.1
2800	2.0
3000	3.0
3300	5.1
3500	7.1
4000	14.6
4500	22.3
5000 (NOTE 2)	28.7

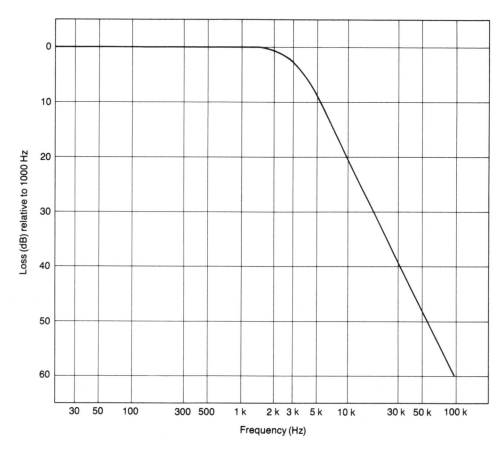

FIGURE 3-2 3-kHz Flat weighting characteristic. (Adapted from ANSI/IEEE Std 743-1984, *IEEE Standard Methods and Equipment for Measuring the Transmission Characteristics of Analog Voice Frequency Circuits*. Copyright 1984 by the Institute of Electrical and Electronics Engineers, Inc., by permission of IEEE.[8])

FIGURE 3-1 C-Message weighting characteristic. (Reprinted from ANSI/IEEE Std 743-1984, *IEEE Standard Methods and Equipment for Measuring the Transmission Characteristics of Analog Voice Frequency Circuits*. Copyright 1984 by the Institute of Electrical and Electronics Engineers, Inc., by permission of IEEE.[8])

NOTES: 1. To ensure that production models of noise measuring sets will be within tolerances, including variations in components and manufacturing procedures, tolerances one half the values given above are recommended for design purposes.

2. The attenuation shall continue to increase at a rate of not less than 12 dB per octave until it reaches a value of 60 dB.

dBrnC.   When noise is filtered by a C-message weighting function prior to measurement, the resulting value is expressed in the unit *dBrnC* (decibels above reference noise, C-message weighted).

dBrnC0.   The unit *dBrnC0* means "decibels above reference noise, C-message weighted, referred to the zero transmission level point." (See Chapter 2 for a discussion of transmission level point.) Noise expressed in dBrnC0 is a *relative* level, whereas noise expressed in dBrnC is an *absolute* level. Note that the unit dBrnC0 is to dBrnC as the unit dBm0 is to dBm. For example, suppose that the noise measured in a carrier system at a +7 TLP has an absolute value of 29 dBrnC. This measurement could be expressed as 22 dBrnC0.

Additional units.   Additional units of noise can be built on the three basic units described above. These additional units are described in the sections that follow.

Metallic Noise

Metallic noise is measured differentially across the ring and tip conductors. (See Chapter 6 for more on metallic versus longitudinal circuits.) Several variations of metallic noise exist, as described below.

Idle circuit noise.   Idle circuit noise (also called *C-message noise* or *idle channel noise*) is the most common steady noise measurement made on telephone circuits. It is made with no signal or modulation present. [The word *idle* here means that the circuit is quiet (i.e., no talking or data). Idle here does not refer to the busy/idle state of a circuit's signaling. Indeed, *idle* circuit noise is usually measured with the circuit in the off-hook or *busy* signaling state.] Idle circuit noise is most annoying to listeners during quiet pauses in speech. The performance objectives for idle circuit noise are set by evaluating subjective tests on listeners. Idle circuit noise measurements are expressed in units of dBrnC or dBrnC0.

C-notched noise.   C-notched noise is the same as idle circuit noise, with the addition of a 1004-Hz *holding tone* inserted at the input end of the circuit under test. At the output end, the holding tone is notched out before the residual noise signal is measured. A C-message weighting filter is also used. This measurement is expressed in the units *dBrnC-notched* or *dBrnC0-notched*. The receiving-end notch filter is centered at 1010 Hz (Fig. 3-3). The apparent frequency discrepancy between 1004 and 1010 Hz is historical and unimportant since the notch filter is wide enough to attenuate holding tones adequately between 995 and 1025 Hz.

C-notched noise measurements test a circuit's noise performance with a signal present. This is important on circuits that use compandors (compressors/expandors). Such a circuit may exhibit in-use noise performance that differs from its idle circuit noise. The holding tone biases the compressor near its normal operating level. Holding tone levels of −13 dBm0 and −16 dBm0 are used when evaluating circuits for data and voice performance. C-notched noise performance

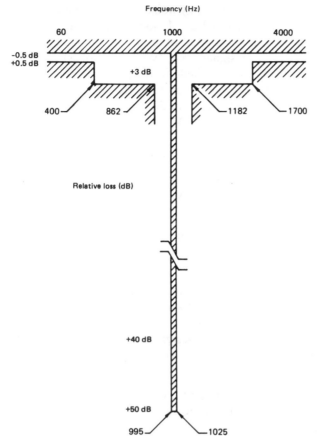

FIGURE 3-3 1010-Hz notch filter. (Reprinted from ANSI/IEEE Std 743-1984, *IEEE Standard Methods and Equipment for Measuring the Transmission Characteristics of Analog Voice Frequency Circuits*. Copyright 1984 by the Institute of Electrical and Electronics Engineers, Inc., by permission of IEEE.[8])

objectives for voice applications are slightly relaxed compared to idle circuit noise objectives. This is allowed by the noise-masking effect of a voice signal being present. C-notched noise performance objectives for data applications are determined by modem error rate requirements.

3-kHz flat noise. This noise measurement is similar to idle circuit noise, but the frequency weighting curve includes 60 Hz and its harmonics. A high metallic 3-kHz flat noise measurement may indicate an unbalanced equipment termination or cable pair. The unit for this measurement is *dBrn 3-kHz flat.*

Quantizing noise. Quantizing noise or *quantizing distortion* is created in digitized speech systems as described above under noise sources. Quantizing distortion will affect a C-notched noise measurement (already described). In this book, when we seek to measure quantizing noise as an isolated impairment, we call it quantizing distortion (see Chapter 7).

Single-frequency noise.   Single-frequency noise is usually measured with a *Frequency-Selective Voltmeter* (FSVM). A FSVM (sometimes called a *wave analyzer*) can be modeled as a voltmeter preceded by a narrow, tunable notch-pass filter. The instrument indicates both signal level (volts or power) and the frequency to which the notch filter is tuned. Single-frequency noise measurements are usually made to investigate the source of noise interference. For this reason the FSVM's frequency indication may be of more diagnostic value than its amplitude reading. Figure 3-15 shows a typical FSVM.

Longitudinal Noise

Longitudinal noise is also called *noise-to-ground* or *power influence* (see Chapter 6). A longitudinal signal is usually measured as the average of the tip-to-ground and ring-to-ground voltages.

C-message (longitudinal).   High longitudinal C-message weighted noise is a potential problem since it may be converted into metallic (and audible) noise by circuit unbalance.

3-kHz flat (longitudinal).   A sufficiently high longitudinal 3-kHz flat noise voltage on a telephone line can be a safety hazard. That is, the 60- and 180-Hz induced power-line voltage may be high enough to create an electric shock. Note that with well-balanced circuits and carefully designed equipment, the circuit may sound and function fine in the presence of dangerously high longitudinal voltages. The longitudinal 3-kHz flat noise measurement is a check for this condition.

Single-frequency (longitudinal).   Single-frequency noise measurements are made longitudinally for the same reason that they are made metallically—to investigate the *source* of excessive noise.

3-1-3 Signal-to-Noise Ratio

Signal-to-Noise (S/N) ratio measurements are not usually made to assess the voice performance of telephone circuits. Some of the reasons follow:

1. Talker levels vary widely, so the signal (and thus signal-to-noise ratio) would vary from call to call—even on the same circuit
2. The presence of a signal (voice) tends to mask noise; the absolute noise level during quiet passages is more important than the S/N ratio during speech
3. Listener assessment is not a direct function of S/N ratio; for example, S/N ratio improvement above 40 dB is usually not subjectively important

Signal-to-noise ratio measurements are more meaningful when made to assess the performance of circuits for voice-band data. This is because the data signal's amplitude is much more predictable than that of speech.

   A signal-to-noise measurement is made in essentially the same way as a C-notched noise measurement. The S/N ratio is simply the ratio between the *received* holding tone level and the measured noise.

3-1-4 Steady Noise Measuring Sets

The instrument that is used to measure steady noise is called a *Noise Measuring Set* (NMS). It is rare, however, to find a modern instrument dedicated to noise measurement. The steady noise measuring function is usually combined with impulse noise, level, and frequency measurements in a multipurpose instrument such as that shown in Fig. 2-2.

Block Diagram

   Figure 3-4 is a block diagram of the noise measuring portion of a typical transmission measuring set. The measuring circuit is ac-connected via large coupling capacitors. The holding coil may be switched in to hold any circuit requiring dc loop current for proper operation. The coil's large inductance has a minimal effect on accuracy. Precision 600- and 900-Ω terminating resistors are provided so that a terminating measurement may be made. The termination is switched out to make a bridging measurement. To indicate the correct noise power, the instrument's internal gain must change when the impedance is changed. The dotted line in the diagram reminds us of this. (Note that the gain change applies to both terminating and bridging measurements.)
   The circuitry discussed so far would be shared in a multifunction instrument. More specific to the noise measuring functions are a series of filters. The 1010-Hz notch reject filter is switched in when making C-notched noise measurements. Next, the appropriate weighting filter (C-message or 3-kHz flat) is selected. The resulting voltage is measured by an rms or quasi-rms detector. (A quasi-rms detector provides a sufficiently accurate rms measurement for waveforms of inter-

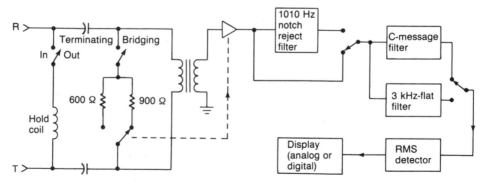

FIGURE 3-4   Block diagram: noise measuring set.

est.) Finally, the signal is displayed on an analog meter or digital display. The quantity displayed is the noise power in dBrn that is dissipated in the 600- or 900-Ω termination.

Steady Noise Measuring Set Standards

In its Standard 743-1984,[8] the Institute of Electrical and Electronics Engineers (IEEE) specifies the features and performance of noise measuring sets. This document is highly recommended for designers of test instruments. Table 3-1 summarizes the input impedance, measurement range, and accuracy specified by the standard. The standard also covers the frequency response accuracy for the weighting and notch filters.

TABLE 3-1   NOISE MEASURING SET STANDARDS

Adapted from ANSI/IEEE Std 743-1984, *IEEE Standard Methods and Equipment for Measuring the Transmission Characteristics of Analog Voice Frequency Circuits*[8]			
		Metallic	Noise to ground
Measurement range	C-message 3-kHz flat C-notched	10–60 dBrnC 20–70 dBrn 20–70 dBrnC	40–110 dBrnC 50–130 dBrn —
Accuracy		±1 dB	±1.5 dB
Input impedance		600 or 900 Ω terminating, ≥20 kΩ bridging	Minimum of 100 kΩ at 20–800 Hz, decreasing to 20 kΩ minimum at 4 kHz[a]

[a] Noise to ground is measured across a high input impedance; however, the quantity displayed is the power that would be dissipated in 600 Ω by the measured voltage.

3-1-5 Measuring Steady Noise

In this section we illustrate the measurement of steady noise using the examples of subscriber loops and digital carrier.

Example 3-1

Several circuit configurations exist for measuring subscriber loop noise. Figure 3-5 shows one method that can be carried out by one person. The measuring set is connected to ring and tip at the subscriber end and is configured for a 600-Ω terminating measurement with the holding coil switched in. A ground connection is also required if noise to ground is to be measured.

If the measuring set provides a means to switch in a butt-in test set, this is done and a CO quiet termination is dialed. (Alternatively, the butt-in may be temporarily

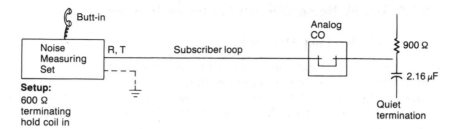

FIGURE 3-5 Noise measurement: subscriber loop.

connected directly to ring and tip. Even simpler is use of a measuring set that contains the butt-in function.) Upon receipt of the digits, an analog CO will switch the connection through to a 900-Ω quiet termination. (A digital CO will connect an idle code to the loop and there will be no through-office connection.)

Next the butt-in is switched out (or removed) and the measurements made—either metallic or noise to ground; either C-message or 3-kHz flat weighting. Note that this measurement includes any office noise (attenuated by the loop loss). Also note that a 600-Ω impedance is used at the subscriber end even though the CO has a nominal 900-Ω impedance.

Example 3-2

Figure 3-6 shows the arrangement for measuring the noise on a channel between the two D-type channel banks of a T-carrier system. This example shows two-wire channel units at each end. We are measuring the noise at the east end, which has a receive Transmission Level Point (TLP) of -2 dB. The NMS and the west-end terminations are set to match their associated channel unit impedance. The termination at the west end is important. Without it, noise generated in the east-to-west direction would reflect from the west end, return to the east end, and falsely increase the noise reading.

Let us assume that the NMS indicates 20 dBrnC. Since we are measuring at a -2-dB transmission level point, this reading is equivalent to 22 dBrnC0. This just meets the C-message noise limit for a D-bank of 23 dBrnC0.

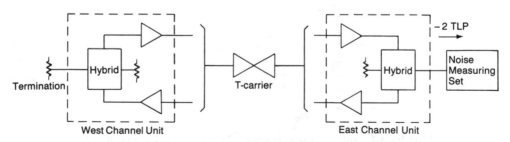

FIGURE 3-6 Idle circuit noise measurement: D-channel bank.

3-1-6 Steady Noise Objectives and Performance

Subscriber Loop Objectives

In the past, steady noise objectives for subscriber loops have been published by the Bell System, independent telephone companies, and the Rural Electrification Administration (REA). These objectives are now covered in a relatively new document—ANSI/IEEE Standard 820-1984, *IEEE Standard Telephone Loop Performance Characteristics.*[9] The limits set on metallic idle circuit (C-message) noise are based on subjective studies of audible noise annoyance. The limits set on metallic 3-kHz flat noise are based on proper equipment operation. Two factors combine to influence the limits set on longitudinal C-message noise: the objective for metallic C-message noise, and the typical balance of cable pairs. Safety considerations determine the limit for longitudinal 3-kHz flat noise. (If the objective of 126 dBrn 3-kHz flat is exceeded, the cable pair may exceed 50 V to ground—the safe limit.)

Table 3-2 at the end of this chapter summarizes the steady noise objectives set by the industry for subscriber loops.

Subscriber Loop Performance

The best, recently published data on subscriber loop performance comes from the 1980 Bell System Noise Survey.[21] The 1980 Noise Survey provides noise data taken under a variety of conditions: at either end of the loop, with and without the CO attached, and both on-hook and off-hook.

Noise measured at subscriber.   The survey configuration that corresponds closest to IEEE Standard 820 is shown in Fig. 3-7. In this configuration, 90% of loops had idle circuit noise of 13 dBrnC or less. The recommended noise limit of 20 dBrnC was met by 96% of loops and the acceptable noise limit of 30 dBrnC was met by 99% of loops. The remaining 1% of loops would be considered unacceptably noisy. The worst loop measured 69 dBrnC.

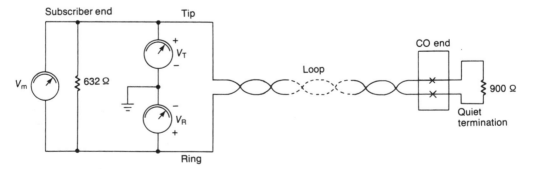

FIGURE 3-7   Noise measured at subscriber end of loop. (Reprinted with permission from AT&T. Copyright 1984 AT&T.)

When longitudinal C-message noise was measured in the same configuration, 90% of loops met the IEEE Standard 820 recommended limit of 80 dBrnC. Also, 98.5% of loops met the acceptable limit of 90 dBrnC. The worst loop measured 101 dBrnC.

Longitudinal noise with 3-kHz flat weighting was also measured in this configuration. Here, 95% of loops measured 110 dBrn or less. One loop, however, exceeded the IEEE Standard 820 safety limit of 126 dBrn. The longitudinal noise on this one loop measured 134 dBrn 3-kHz flat.

Noise measured at CO.   Other circuit configurations used in the 1980 Noise Survey do not correspond to the measurement methods specified in IEEE Standard 820. The survey data are nevertheless summarized below, but we will not be able to compare the data with objectives.

When idle circuit noise was measured at the CO end with an on-hook telephone at the subscriber end (Fig. 3-8a), 95% of loops measured 22 dBrnC or less. This circuit arrangement is significant since it can be used from the CO on an automated basis without field personnel.

When a resistor replaced the on-hook telephone to provide a more realistic circuit (Fig. 3-8b), 95% of loops measured 15 dBrnC or less.

Longitudinal noise (using either C-message or 3-kHz flat weighting) was considerably lower at the CO end (Fig. 3-9) than at the subscriber end. This is probably due to the CO's low longitudinal impedance compared to that of the subscriber end.

Trunk Objectives

Bell Communications Research (Bellcore) publishes trunk noise objectives in its *Notes on the BOC Intra-LATA Networks—1986*.[13] The objectives are a

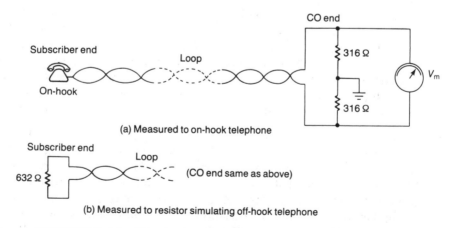

(a) Measured to on-hook telephone

(b) Measured to resistor simulating off-hook telephone

FIGURE 3-8   Metallic noise measured at CO end of loop. (Reprinted with permission from AT&T. Copyright 1984 AT&T.)

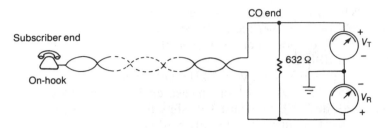

FIGURE 3-9   Longitudinal noise measured at CO end of loop. (Reprinted with permission from AT&T. Copyright 1984 AT&T.)

function of the facility used to implement the trunk. For trunks on an analog carrier system, the noise objectives are also a function of trunk length. Table 3-3 shows Bellcore's objectives for steady noise on trunks.

Trunk Performance

We do not have data on the steady noise performance of isolated trunks; instead, we report the noise performance of trunks connected in tandem. These data are from the Bell System 1982/83 End Office Connection Study (EOCS).[22] The EOCS tested Bell's nationwide trunk network by making toll calls from end office (local CO) to end office. No subscriber loops were connected. In the EOCS, connections were categorized by airline miles as short (0 to 180 mi), medium (181 to 720 mi), and long (721 to 2576 mi). A summary of the EOCS steady noise data follows.

Idle circuit noise.   For short connections, 99% of calls had noise between 12 and 36 dBrnC. Virtually no calls were better than 12 dBrnC and the remaining 1% of short connections exceeded 36 dBrnC.

For medium-distance and long connections, 98% of calls had noise between 23 and 42 dBrnC. One percent of calls were better than 23 dBrnC and 1% were worse than 42 dBrnC. Some calls were as bad as 60 dBrnC. Idle circuit noise correlated closely with call distance; the greater the distance, the higher the noise.

The idle circuit noise performance of trunks in tandem from the EOCS agrees fairly well with the current objectives for single trunks (Table 3-3) if you consider that trunks in tandem will have more noise than will isolated trunks.

C-notched noise.   For short connections, 1% were better than 22 dBrnC and 1% were worse than 46 dBrnC. The remaining 98% of calls fell in the middle. For medium-distance and long connections, the first and 99th percentiles were 32 and 46 dBrnC. C-notched noise was less a function of distance than was idle circuit noise. This is probably due to the existence of at least one T-carrier link (usually the tandem connecting trunk) in each built-up connection. Any call with one (or more) T-carrier links will have a minimum C-notched noise (independent of distance) due to the quantizing distortion of the T-carrier channel.

3-kHz flat noise.   For calls of any distance, the first and 99th percentiles were 27 and 60 dBrn 3-kHz flat. The 3-kHz flat noise of toll calls was not closely related to distance. This is because 3-kHz flat noise is generally caused by 60-Hz influence in the *local* trunk plant.

Signal-to-Noise ratio.   In the EOCS, S/N ratio was measured by placing a -12-dBm sine wave at 1004 Hz at one end of the connection. At the other end, the received 1004-Hz level was measured. The 1004 Hz was then notched out and the residual noise was measured via a C-message weighting filter. The S/N ratio was the ratio between the received tone and the C-notched noise. Ninety-five percent of calls fell within an S/N ratio range of 30 to 40 dB. Virtually no calls had an S/N ratio above 40 dB. This is probably due to the quantizing noise of at least one T-carrier link in the connection, as explained above. Almost all the remaining 5% of calls had an S/N ratio of 15 dB or better.

For reasons discussed earlier in this chapter, the S/N performance reported here has more bearing on modem performance than on voice performance. Modem designers should find the data presented in the EOCS of great value. For example, even the summary of data presented here implies that modems should function with S/N ratios as poor as 15 dB.

Longitudinal noise.   Longitudinal noise on toll calls is a local phenomenon that depends on office location and not on call distance. The worst readings were 85 dBrn—much better than the worst loops from the 1980 Noise Survey.

Carrier System Objectives

The steady noise objectives for digital carrier systems are listed in Table 3-4. Note that there is a 5-dB margin between the end-to-end idle circuit noise requirement for a digital channel and the preservice limit for a complete trunk based on that channel (23 dBrnC0 versus 28 dBrnC0).

Switching Office Objectives

The quantity measured to characterize the noise performance of switching equipment is called *cross-office noise*. It is measured by establishing a connection through the office to a test termination. Noise is then measured at the originating end using a noise measuring set. The NMS and the far-end termination should each provide dc paths. This will allow loop current to both hold the connection and put the office equipment in its normal operating state.

Cross-office noise varies from connection to connection. Many connections made at random are usually set up to obtain a statistically valid sample. Table 3-5 lists the steady noise objectives for various office types. Note that digital offices are granted higher steady noise limits than analog offices, to allow for the A/D converter noise of the digital office.

Switching Office Performance

We do not have steady noise performance for switching offices as stand-alone entities. Currently, the noise on built-up connections is predominated by trunk and loop noise; cross-office noise is usually not significant. However, as the network evolves toward a totally digital implementation, this will change. On a future built-up connection, the only noise sources will be one digital office at each end. The interconnecting digital trunks and offices will be virtually noiseless.

Subscriber Equipment Objectives

The Electronics Industries Association (EIA) publishes a series of Recommended Standards for subscriber equipment such as Private Branch Exchanges (PBXs), key systems, and individual telephone sets. The steady noise objectives from these documents are summarized in Table 3-6. These standards recommend that subscriber equipment contribute little noise to a connection.

Subscriber Equipment Performance

Unless it is defective, subscriber equipment probably contributes little to the total noise on a connection. However, it is all too easy for defective subscriber equipment to exist. In many cases, the local telephone company no longer maintains residential equipment. The EIA standards are recommendations only. In the United States, the only *requirements* on subscriber equipment are set by the FCC and UL. These organizations are concerned mainly with safety, however, not with performance. If you drop a cheaply constructed phone on the floor, you may find yourself talking over the static resulting from internal damage. Careful planning and design by the telephone companies and interexchange carriers can provide low-noise circuits in most of the network. However, subscriber equipment is out of their control and is thus a weak link.

Total Network Objectives

The steady noise objectives for the total telephone network (from subscriber to subscriber) are set by making subjective surveys. In general, the user's expectation is that noise will increase with connection distance. For voice performance, the objectives are summarized in Table 3-7. For data performance, Bellcore has set an intra-LATA signal-to-noise objective: The ratio of the received tone to the C-notched noise should be 24 dB minimum, subscriber to subscriber.

Total Network Performance

We have previously cited noise performance data from two separate surveys: a survey of toll calls between end offices, and a survey of subscriber loops. To see the total picture (subscriber to subscriber), we must combine these two groups of data. The authors of the EOCS have done this.

When the loop performance was added to the toll performance, the resulting subscriber-to-subscriber noise was less than the end office-to-end office noise alone. In other words, the loops' losses tended to attenuate the noise from the toll network without adding significant noise of their own. To better appreciate this, consider the following. The median toll call had an end office-to-end office noise of 24 dBrnC. The median loop had a subscriber-to-CO noise of less than 0 dBrnC and a loss of 3.5 dB. Clearly, if we add this loop to the median toll call, the loop would add nothing to the total connection noise. The loop loss would, however, attenuate the noise from the toll network to 20.5 dBrnC at the subscriber end.

When the toll and loop noise distributions were added statistically, 95% of long-haul toll calls had a subscriber-to-subscriber performance of 34 dBrnC or better. This performance is in approximate compliance with the objective stated in Table 3-7 (for a long connection with average noise plus two standard deviations).

3-2 IMPULSE NOISE

3-2-1 General Characteristics

At high amplitudes, impulse noise is heard as clicks or pops. At lower amplitudes, impulse noise may be inaudible, yet still cause problems with voice-band data transmission. This is in contrast to steady noise, which usually causes listener annoyance first, before data errors, as its level increases.

The general nature of impulse noise is shown in Fig. 3-10. We have shown

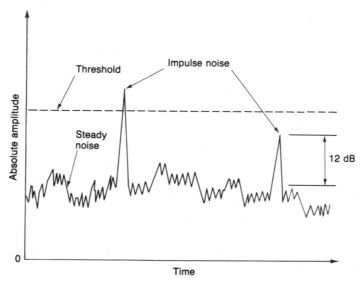

FIGURE 3-10   Impulse noise.

absolute (rectified) amplitudes; the actual ac signal on a circuit would have both positive and negative impulses. A pulse that is 12 dB greater than the rms steady noise is usually considered to be impulse noise. Pulses below this level are merely peaks of the steady noise. Impulse noise transients are counted when they exceed a specified threshold. Figure 3-10 shows one such pulse. Impulse noise is loosely defined as the threshold at which the count is an average of one per minute. (Actual specifications are more complex and specify 5- or 15-minute counting intervals.) Impulse noise level objectives lie in the region above steady noise objectives and below data signal levels.

Sources Of Impulse Noise

Intermittent connections.   Bad splices and corroded connections can cause impulse noise. Consider that a subscriber loop is fed from 48 V and passes a current of 20 mA dc or greater. These dc signals are huge compared to the minute ac levels of voice and data. A slight but sudden resistance change (caused by a poor connection) will cause a large transient in the ac signal. A demonstration of this phenomenon can be heard by using a telephone with a bad cord.

Relay contacts.   Another source of impulse noise is arcing in relay contacts. For example, a relay may open to break the flow of ringing current to a subscriber loop. The combination of high voltages, high currents, and reactive loads may cause contact arcing. Energy may radiate into adjacent low-level circuitry and cause impulse noise.

Miscellaneous impulses.   Other sources of impulse noise include very high 60-Hz induction and crosstalk from other circuits. For a while, one of the most irritating, data-disrupting transients was the call-waiting signal. New CO generic software now allows call-waiting to be canceled on a per-call basis. This is a boon for users who want their call-waiting and data, too!

Impulse Noise Reduction

The mitigation of impulse noise takes place both during equipment design and at system layout.

Equipment design.   The increased use of solid-state devices in lieu of relays significantly reduces impulse noise. Those relays that remain should be equipped with contact arc suppression devices. Another useful technique (especially for ringing relays) is *zero-crossing switching*. Here the circuit is designed so that relay contacts close when the switched voltage is zero and open when the current is zero. The triac is a useful solid-state device (and relay replacement) that inherently opens its load circuit at the zero-current crossing. Careful PCB layout, power supply filtering, and cable dress are also important in equipment design to reduce impulse noise.

System layout.   Impulse noise on carrier systems can be reduced by using a short repeater section at the CO. Since CO switching equipment is a source of impulse noise, a short CO repeater section will allow a higher carrier-to-impulse noise ratio at the CO. This will result in lower impulse noise in the demodulated voice channels. Another noise reduction technique is to assign VF subscriber pairs and carrier system pairs to separately shielded cables. Single cables with two separately shielded sections are available and could be used here.

In step-by-step offices, the stepping switches must be properly maintained (lubricated and adjusted) to reduce this source of impulse noise.

Measurement Units

Impulse noise is measured in *counts* per unit of time. A count is an excursion of the noise waveform above a specified *threshold* level. The noise is usually frequency weighted (C-message or 3-kHz flat) before the impulses are counted. Impulse noise is sometimes measured using a holding tone and C-notched weighting. A typical impulse noise specification might read "10 counts or less above 41 dBrnC0 in 15 minutes."

Individual circuits are usually measured for 15 minutes. If a group of circuits sharing the same facility is to be evaluated, each circuit may be measured for only 5 minutes. It is also possible to use a sampling technique in lieu of measuring every circuit in a group. In its specifications, Bellcore describes a *sequential sampling* scheme. This scheme allows a circuit group with good performance to be accepted quickly by measuring only a few circuits (and similarly, a group with poor performance to be rejected quickly). Testing of additional circuits is required only if the group shows marginal performance. A standard may specify that a group is acceptable if at least 50% of its circuits measure less than a certain count limit. Using a sampling technique, it is often not necessary to measure 50% of the circuits. Note that this streamlined testing for circuits in a group is based in part on the assumption that all the circuits of one group are under a similar noise influence.

3-2-2 Measuring Instruments for Impulse Noise

Figure 3-11 shows the block diagram of a basic impulse noise measuring set. To use the instrument, the operator makes the following adjustments:

1. Input conditions (terminating or bridging, 600- or 900-Ω, metallic or longitudinal)
2. Frequency weighting (C-message, C-notched, 3-kHz flat)
3. Threshold (peak power at the input in dBrn)
4. Counting period (5 minutes, 15 minutes, continuous)

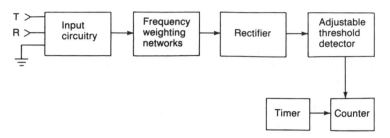

FIGURE 3-11   Block diagram: impulse noise measuring set.

At the end of the counting period, the counter will display the number of times that noise impulses exceeded the threshold. Frequently, three thresholds and three counters are provided so that a rough impulse noise distribution can be determined. In Chapter 7 we discuss impulse noise measurement in conjunction with two other transients—hits and dropout.

The impulse noise measuring function may be included in a multipurpose instrument such as that shown in Fig. 2-2. In this case, some of the elements shown in the block diagram would be shared by other measurement functions.

Impulse Noise Measuring Set Standards

The IEEE's Standard 743-1984[8] contains detailed specifications for impulse noise measuring sets. A summary of these specifications follows.

Threshold.   For metallic measurements, the threshold must be settable within the range 30 to 110 dBrn with an accuracy of ±1 dB. For longitudinal measurements, the range is 60 to 140 dBrn and the required accuracy is ±1.5 dB. Note that the threshold is in terms of peak power at the instrument's input.

Filters.   The C-message and 1010-Hz notch filters used for impulse noise have the same frequency response as the filters used for steady noise. However, Standard 743-1984 also specifies the poles and zeros for these filters when they are used to frequency weight impulse noise. This is because the pole and zero locations affect the filter's response to impulses.

Counters.   The counters must be able to register at least 998 counts and include an overflow indication. The counters must be limited to a maximum count rate of 8 per second so that ringing in the filter circuitry will not cause extraneous counts.

Calibration.   The calibration of an impulse noise measuring set is checked by switching in both the C-message and 1010-Hz notch filters and applying an 1800-Hz sine wave at 0 dBm to the input. There should be no counts when the threshold is set at 92 dBrn. The counter should count continuously when the threshold is set at 91 dBrn.

3-2-3 Measuring Impulse Noise

In this section we illustrate the measurement of impulse noise at both the system and equipment levels.

Example 3-3

Figure 3-12 shows a typical instrument arrangement for measuring impulse noise performance at the system level. We have selected an electronic CO for the example. The control settings for the measuring set are shown in the figure. Lines A and B are selected according to a sampling scheme. For each randomly selected pair of lines, a call is established through the switch and the impulses are counted for 5 minutes. The CO meets the performance objective if 50% of the calls have counts not exceeding five in 5 minutes above 47 dBrnC (Table 3-11).

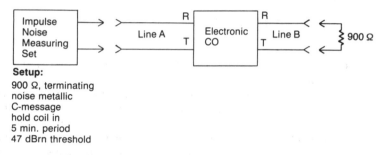

FIGURE 3-12   Typical impulse noise measurement: system level.

System-level tests are performed on equipment that is operating in its typical impulse noise environment. For a CO, this means a switch that is carrying either live or simulated traffic. The test results are dependent on several variables: hardware design, office battery voltage, traffic level, ringing load, and loop conditions. Recall that the system-level objective allows some connections to measure *above* the count limit and expects others to be *below* the limit. To the engineer designing an individual part of the total system, this objective may appear to be a moving target. In the next example, we modify the system-level objective to apply to equipment-level design.

Example 3-4

The approach we recommend for testing impulse noise at the equipment level is to analyze your design for worst-case impulse noise scenarios. Continuing with the electronic CO from Example 3-3, we illustrate this approach in Fig. 3-13. Assume that we have a digital CO with per-line codecs. This allows the simplifying assumption that all impulse noise is generated in the analog portion of the line cards. We then analyze the line cards for possible impulse noise sources. Some that we find are dial pulsing, ringing, ringing interruption, and ring trip. These activities on one line can cause impulse noise on another line. The coupling is often between adjacent circuits on one PCB or between circuits on cards in adjacent slots of a card cage.

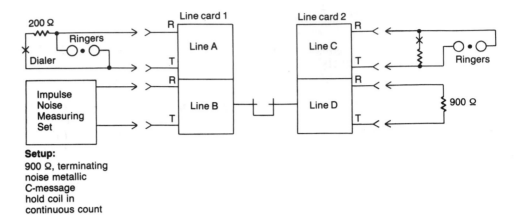

Setup:
900 Ω, terminating
noise metallic
C-message
hold coil in
continuous count

FIGURE 3-13 Typical impulse noise measurement: design stage.

Coupling can also take place between physically separate circuits that share a common power supply.

In Fig. 3-13 we show how to investigate interference between adjacent circuits on the same PCB. A source of impulse noise is provided by sending continuous dial pulses into line A. The dial pulse transients on this line are made worse by electrically simulating a short loop with a heavy ringer load. Next place a call between line B and line D. Connect an idle termination and impulse noise measuring set as shown. Adjust the measuring set threshold downward until the counter starts to count regularly. This threshold will probably be sharply defined due to the repetitive nature of the impulse source. If this threshold is not at least 10 dB below the objective threshold, the line card coupling should be found and reduced.

During this investigation, it might be helpful to replace the impulse NMS with an oscilloscope whose sweep is synchronized to the dialer. Now observe the offending impulse on the oscilloscope while trying various fixes on the line card. Try bending suspected components apart or putting small pieces of electrostatic or magnetic shielding between them. Try adding extra short-leaded filter capacitors to appropriate power and ground points. You can also use signal-tracking techniques by moving the oscilloscope from the ring and tip to other points on the PCB in an attempt to see exactly where the impulses enter. Even though your first-pass PCB layout was carefully planned, do not be surprised if it takes a lot of bench time and a second PCB pass to obtain a quiet line card, free of impulse noise.

Referring again to Fig. 3-13, there are additional worst-case configurations to investigate. For one, exchange the measuring set and termination on lines B and D. This will check for coupling from line A to the other direction of line B's transmission circuit. Now apply interrupted ringing to line C and check for impulse noise on lines B and D. Attach a heavy ringer load to line C to make things worse. Trip ringing on line C and see if that causes impulses anywhere. In your switch architecture there may be additional sources of impulse noise. Think this through and set up experiments similar to that of Fig. 3-13 to check for additional problems. This is an area where the published standards are of little help. The hardware designer must use knowledge of the specific system to develop tests. If you do a good job here, your design will easily pass a system-level test.

Here is a word of caution about benchtop experiments of the type shown in Fig. 3-13. The transients generated by the dialer contacts on line A and the hook switch contacts on line C can be quite severe. These transients may radiate directly into the line card under test, bypassing its tip and ring conductors. The transients may also radiate directly into the NMS, bypassing the equipment under test altogether. Either condition will result in erroneous test results. If you suspect you have this problem, locate the dialers and ringers associated with lines A and C 25 to 50 feet away from the other equipment. This is a valid simulation of field conditions since even the shortest telephone loop has 25 or 50 feet of cable between the line circuit and the nearest telephone.

3-2-4 Impulse Noise Objectives and Performance

Impulse noise objectives are based on the data performance of voice circuits.

Subscriber Loops

Bellcore specifies limits for impulse noise on subscriber loops, as shown in Table 3-8.

Trunks

Objectives.   Bellcore specifies limits for impulse noise on trunks in intra-LATA networks (Table 3-9). The objectives are a function of the facility type that is used for the trunk.

Performance.   The 1982/83 End Office Connection Study[22] showed that impulse noise performance is not a function of trunk length, but rather, is a function of end office type. The EOCS data are presented later under "Switching Equipment."

Carrier Systems

Impulse noise objectives for carrier systems are summarized in Table 3-10.

Switching Equipment

Objectives.   Objectives for various switching office types are listed in Table 3-11. Note that the newer technology offices are allowed less impulse noise.

Performance.   The EOCS included the measurement of impulse noise on the built-up connections. Impulse noise was found not to correlate well with distance. Impulse noise was, however, closely related to end office type. We will thus assume that most of the measured impulse noise was due to the switching equipment.

The thresholds used in the EOCS were relative to a received holding tone. The lowest threshold was 12 dB below the holding tone. Since the holding tone was transmitted at -12 dBm and the mean connection loss was 7 dB, the relative threshold corresponds roughly to an absolute threshold of 59 dBrnC.

In the EOCS, 80% of calls between electronic offices had five or fewer impulse noise counts in 5 minutes. For calls between step-by-step offices, the performance degrades to 35% of calls having five or fewer counts in 5 minutes. The performance for crossbar equipment fell between that of electronic and step-by-step offices. As we would expect, impulse noise was found to increase during the busy hour, especially for step-by-step offices.

Due mainly to the different thresholds used, it is not possible to comment on the EOCS performance versus the objectives of Table 3-11. Nevertheless, the impulse noise data presented in the EOCS may be of use to designers of modems and other persons interested in the data performance of the network.

Subscriber Equipment

Impulse noise objectives for subscriber equipment are summarized in Table 3-12. Regarding impulse noise performance of subscriber equipment, our earlier comments about steady noise performance apply here also.

Complete Network

Bellcore has specified an impulse noise objective for intra-LATA networks (Table 3-13). A rough interpretation of the EOCS data shows that the objective is met for calls between electronic offices. For calls between crossbar or step-by-step offices, performance falls just outside the objective.

3-3 CROSSTALK

3-3-1 General Characteristics

Crosstalk is simply a desired signal found in the wrong place. In this section we classify crosstalk as *unintelligible* or *intelligible*. We then list sources of crosstalk and ways to reduce this impairment. Finally, we present industry objectives for crosstalk performance.

Unintelligible Crosstalk

Crosstalk may be unintelligible due to its low level or due to a frequency conversion occurring during the crosstalk process. Unintelligible crosstalk may sound like babble. At the equipment and design levels, it may be desirable to investigate individual sources of unintelligible crosstalk. However, at the system or network level, the babble of unintelligible crosstalk is usually lumped with steady noise measurements. Note that unintelligible crosstalk can be objectionable to users since it often sounds like it might be intelligible.

Intelligible Crosstalk

Intelligible crosstalk is more objectionable than most other impairments since it results in a loss of privacy. The listener hears another caller's conversation and perceives that his own conversation may also be overheard. Trunks are allowed to have more frequent occurrences of intelligible crosstalk than are loops. This is because, in general, callers on trunks are unknown to each other and are randomly paired by the switching process. On the other hand, callers on local loops may be known to each other. Also, subscriber assignments on local cable pairs are fixed.

There is no standard method for measuring intelligible crosstalk as separate from unintelligible crosstalk; however, there are objectives for incidences of intelligible crosstalk, as described later.

Sources of Crosstalk

Many sources of crosstalk are due to electromagnetic coupling between physically adjacent circuits. In this subsection we briefly describe these and other sources of crosstalk.

Near- and far-end crosstalk in cables. Near-End Crosstalk (NEXT) and *Far-End Crosstalk* (FEXT) occur between cable pairs (Fig. 3-14). Combinations of NEXT and FEXT are possible via tertiary pairs.

The first step toward achieving low crosstalk in cable is to use high-quality cable that has been carefully manufactured to assure high balance. The REA and others provide help by listing quality cable on their lists of approved material. The second step is proper installation—specifically, correct splicing technique as recommended by the manufacturer of the splicing connectors. Third, newly installed cable should undergo acceptance tests to verify its compliance with crosstalk and balance objectives.

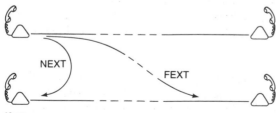

Key:
NEXT Near-End Crosstalk
FEXT Far-End Crosstalk

FIGURE 3-14 Crosstalk on cable pairs.

Circuit unbalance. Poor metallic-to-longitudinal balance can be a source of crosstalk coupling. This phenomenon is described in Chapter 6. Terminal equipment designed for good balance is the answer to this problem.

Components and circuit boards. Within equipment, undesired coupling often occurs on a printed circuit board as capacitive coupling between traces or components. Large components from adjacent circuits should be physically separated. The outside foil of wrapped capacitors may be connected to circuit ground (or circuit low point) so as to act as a shield. Careful trace layout is essential. It may be necessary to use shielding ground traces between adjacent circuits.

Transformers are another source of crosstalk coupling. Within a PCB, it may be possible to orient transformers to null interfering fields and thus prevent crosstalk. Between adjacent boards in a card cage this is not possible; the transformers on adjacent boards will have identical orientations. Here it may be necessary to use magnetic shielding between units.

Shared codecs. In early digital carrier channel banks (such as the D2 and D3), one or two codecs are shared by all the channels in the bank. The codec (coder/decoder) performs the analog-to-digital conversion in these systems. To allow many channels to time share a single codec, an analog time-division matrix is placed between the channels and the codec. In its simplest form, this matrix is often a single backplane wire or trace onto which Pulse-Amplitude-Modulated (PAM) voice samples are gated by analog switches. A vestige of the previous PAM sample will remain on the bus during the next time slot, creating a source of crosstalk. Analog circuit design skills are required to keep this crosstalk to an acceptable minimum.

For newer designs, the availability of low-cost, single-chip codecs allows each channel unit to have its own coder. The descriptive jargon here is "per-channel codec." These designs eliminate the PAM bus as a source of crosstalk since all multiplexing is done at logic levels on already digitized voice samples.

Excessive repeater gain. Another source of crosstalk at the system level is use of excessive repeater gain on cable pairs. No matter how well manufactured and installed, there is still some coupling between cable pairs. Thus there must be coordination of the absolute levels of signals placed on pairs in the same cable. One would not combine a microvolt system with one operating at tens of volts. The latter would always be crosstalking into the former.

The level coordination that exists on voice-frequency cable pairs can be upset if too much repeater gain is placed at a single location along a circuit.

Example 3-5

Take the case of an exchange cable that serves subscribers both near and distant. The subscribers near the CO need no amplification. However, imagine one subscriber many miles distant from the CO who needs 20 dB of gain to overcome cable loss and enjoy a satisfactory circuit. One solution might be to place a 20-dB gain repeater on this subscriber's pair at the CO. Consider, however, the relative levels in the cable as it leaves the CO. Parties talking to our distant subscriber will be 20 dB hotter than other calls in the cable. The crosstalk performance would be degraded by 20 dB, possibly becoming objectionable. This is an example of excessive repeater gain causing crosstalk. (Such a high gain may also contribute to repeater instability since the CO side of the repeater may have to balance against changing impedances from call to call.)

   A better solution is to provide 8 dB of gain at the CO and 12 dB of gain at a repeater installed in the field between the CO and the subscriber. A typical rule of thumb for such circuits is 8 dB maximum repeater gain at the CO and 12 dB maximum repeater gain at any one field location. The higher field gain is allowed since signals coming from either direction will be attenuated by a length of cable. (Note that since field-mounted equipment is now involved, it may be cost-effective to serve this subscriber on carrier—and perhaps provide a better-quality circuit.)

   PCM crosstalk floor.   In PCM systems, there is a crosstalk floor that results from a phenomenon similar to that which produces the noise floor. Assume a PCM coder biased just at a decision point. Any source of analog crosstalk present will cause the coder output LSB to change between 0 and 1 in step with the interfering signal. At the decoder end, the resulting analog output will transit a full coder step in response to what might have been a miniscule signal at the encoder end. In summary, the crosstalk has been nonlinearly amplified. The level of this crosstalk floor is a function of system parameters. In practice, the crosstalk floor is masked and randomized by the noise floor so that intelligible crosstalk usually does not occur.[16,26]

3-3-2 Crosstalk Measurement

Test Instruments

   We know of no instrument that has been designed as a dedicated crosstalk measuring set. The instruments used instead are simply an oscillator and a Frequency-Selective Voltmeter (FSVM). The frequency-selective characteristic of the FSVM allows very low crosstalk components to be read in the presence of noise. (FSVMs were discussed earlier under the subject of single-frequency noise.) When measuring crosstalk, the FSVM is usually arranged to measure the voltage across a terminating resistor and to display the results as a power. Figure 3-15 shows a typical FSVM.

Measurement Method

   Technical documents usually specify *equal-level* crosstalk performance. "Equal level" means equal absolute levels at equal transmission level points.

Example 3-6

   Figure 3-16a shows a measurement example with the question of levels at its simplest. We have a local CO with all ports at 0 TLP. Most crosstalk specifications state

FIGURE 3-15 Frequency selective voltmeter. (Courtesy of Hewlett-Packard Co.)

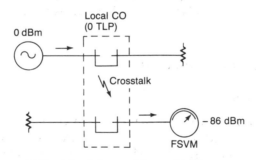

(a) Simple levels (crosstalk loss = 86 dB)

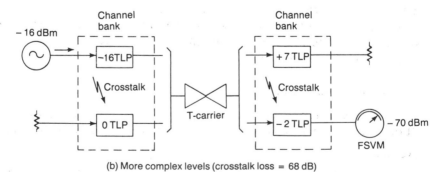

(b) More complex levels (crosstalk loss = 68 dB)

FIGURE 3-16 Examples of equal-level crosstalk measurement.

that the interfering signal be applied at 0 dBm0. Therefore, we insert 0 dBm into one port and measure the crosstalk level at the interfered port. Assume that we measure −86 dBm. With 0 dBm input and −86 dBm output, the crosstalk loss is clearly 86 dB. (Note the use of terminations at the other two ports. Not shown in the figure but usually required by the specification are dc terminations at all four ports.)

Example 3-7

Figure 3-16b shows a somewhat more complicated measurement made between two end-to-end channels of a carrier system. The interfering channel is equipped with four-wire channel units that have a −16 TLP input and a +7 TLP output. The interfered channel is equipped with two-wire channel units having a 0 TLP input and a −2 TLP output.

A level of −16 dBm (0 dBm0) is applied to the interfering channel. The FSVM reads −70 dBm. What is the crosstalk loss? The best way to avoid confusion here is to use the unit dBm0 to express all levels. The FSVM reads −70 dBm at a −2 TLP, which equates to −68 dBm0. A 0-dBm0 input and a −68-dBm0 output give us a crosstalk loss of 68 dB. That is, the equal-level crosstalk loss is 68 dB.

System-Level Measurements

A typical system whose crosstalk performance is under consideration might be a CO or PBX switching machine or a group of trunks carried on the same facility or carrier system. At the system level, it is not possible to measure crosstalk between all combinations of circuits. This is especially true in a switching office where the circuit configuration changes constantly as calls are set up and dropped. Nor is it generally feasible at the system level to anticipate offending circuit paths. If it were, only those paths would need to be measured.

To be practical, crosstalk performance specifications allow measurements to be made using sampling techniques. The specified sampling technique varies from standard to standard; there appears to be no industry-wide standard for sampling. Thus, in the tables at the end of the chapter, we have included the intended sampling method as part of each crosstalk specification.

Equipment-Level Measurements

At the equipment and design levels, it is usually possible (and desirable) to concentrate crosstalk measurements on likely offending circuit paths. For example, the manufacturer of a D-channel bank may guarantee a minimum crosstalk loss between *any* two channels. The manufacturer may also state that to verify the performance, only certain channel combinations need be measured. These combinations would be based on the designer's knowledge of the paths most likely to cause problems. Such combinations include channels sampled sequentially by a shared codec, and adjacent channels on a circuit card.

Example 3-8

Consider the design of a digital PBX using per-channel codecs. Here, the most likely crosstalk paths are between physically adjacent channels—either on the same PCB or between adjacent PCBs. The applicable equipment specification is EIA-464. This standard requires that 95% of cross-office connections have a crosstalk loss of ≥75 dB, and that no connection have crosstalk loss of <70 dB. How does the designer cope with such probability concepts as "95% of connections" when in the project's early stages there may exist only one or two PCBs? In the following, we present a design plan that answers this question.

First, think through the architecture and identify the worst crosstalk paths. We mentioned two above. Second, verify that the equipment design meets the "no connection worse than" limit of 70 dB for these selected paths. Third, assume that when a large system is assembled and many measurements made, the system requirement (95% ≥75 dB) will be met. This assumption is valid since the worst-case paths will be a small part of the total paths randomly sampled at the system level. Most paths through a large system will probably have nonmeasurable crosstalk levels.

Thus the designer should concentrate on a few paths selected as being worst case. Figure 3-17 shows one example. We have devised a worst-case scenario where two cross-office calls appear on adjacent circuits twice (shown as coupling paths 1 and 2). One of our worst-case assumptions is that the relative phases of the two crosstalk sources are such that coupled signals are additive. (They could cancel each other, but don't count on it.) The two crosstalk sources are correlated (since both originate at line A), so they add on a voltage basis. (If a signal's voltage is doubled, the signal's power increases 6 dB.)

Since EIA-464 allows a worst-case crosstalk loss of 70 dB, we must achieve 76-dB isolation for *each* end of the connection. In other words, any single coupling path (such as between circuits on one PCB, or between circuits on adjacent PCBs) must be verified for 76-dB crosstalk loss or better. We have simplified the designer's task. All you need do is test the various individual coupling paths and verify that they

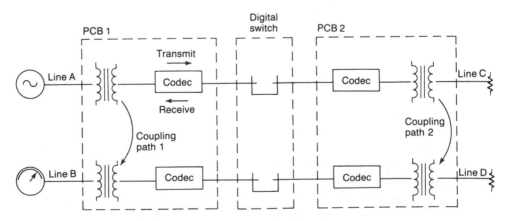

FIGURE 3-17 Worst case crosstalk scenario.

meet 76-dB isolation over the specified frequency range. Note that many of these paths exist within one PCB, so can be verified early in the project (when perhaps only one PCB prototype has been built).

Do not forget to test both directions of transmission on four-wire circuits. On PCB 1 in Fig. 3-17, for example, you must test the following crosstalk paths:

* Line A transmit to line B receive (shown as coupling path 1)
  Line A transmit to line B transmit
  Line A receive to line B receive
  Line A receive to line B transmit
* Line B transmit to line A receive
  Line B transmit to line A transmit
  Line B receive to line A receive
  Line B receive to line A transmit

The two starred paths may be tested using only PCB 1; lines C and D are replaced by idle terminations. The other six paths require that lines C and D be connected so that access is possible to the transmit and receive paths of PCB 1. Since we are now testing individual crosstalk sources on PCB 1, lines C and D should be on different (and separated) PCBs. Otherwise, coupling between lines C and D would spoil the measurement.

In the example, we concentrated on crosstalk coupling that is a function of physical proximity of circuits. There may also be coupling via power supply or other common leads. This coupling might not be related to a circuit's physical location. If you have crosstalk between circuits that is not related to their physical proximity, look for a lack of adequate power supply decoupling.

3-3-3 Crosstalk Objectives

Crosstalk Index

The *crosstalk index* is the probability, expressed in percent, that one or more words from an interfering source will be intelligible during a call. Bellcore has established crosstalk indexes for subscriber loops (Table 3-14) and intra-LATA trunks (Table 3-15). For the reasons explained earlier, better crosstalk performance is required on loops than on trunks.

Other Objectives

Crosstalk index is not readily measured. Standards that call for a single-frequency, equal-level crosstalk measurement are more useful to equipment designers. Fortunately, Bellcore, the REA, and the EIA provide such standards for carrier, switching, and station equipment (Tables 3-16, 3-17, and 3-18).

3-4 REFERENCE TABLES

TABLE 3-2   STEADY NOISE OBJECTIVES: SUBSCRIBER LOOPS

Adapted from ANSI/IEEE Std 820-1984, *IEEE Standard Telephone Loop Performance Characteristics*[9]				
Circuit[a]	Weighting	Recommended	Acceptable	Not acceptable
Metallic	C-message weighted 3-kHz flat	≤20 dBrnC ≤40 dBrn	20–30 dBrnC 40–60 dBrn	>30 dBrnC >60 dBrn
Longitudinal	C-message weighted 3-kHz flat	≤80 dBrnC —	80–90 dBrnC —	>90 dBrnC >126 dBrn

Adapted from Bellcore *Notes on the BOC Intra-LATA Networks—1986*[13]			
Condition	Objective	Marginal	Unacceptable
Metallic, idle circuit	≤20 dBrnC	21–30 dBrnC	>30 dBrnC

[a] Measured at subscriber end of loop with CO end switched through office to a 900-Ω in series with 2.16-μF termination (analog offices), or with CO end switched to idle code (digital offices).

TABLE 3-3 STEADY NOISE OBJECTIVES: TRUNKS

Adapted from Bellcore *Notes on the BOC Intra-LATA Networks—1986* [13]			
Facility type[a]	Distance (miles)	Noise limits[b] (In dBrnC0 unless otherwise stated)	
		Preservice	Immediate action
Noncompandored analog carrier	0–50	31	40
	51–100	33	40
	101–200	35	40
	201–400	37	42
	401–1000	40	44
Compandored analog carrier	0–50	26	34
	51–100	28	34
	101–200	30	34
	201–400	32	36
	401–1000	35	38
Digital carrier	Any	28	34
Voice frequency metallic (TCT with EML > 6 dB)[c]	Any	20 dBrnC	30 dBrnC
Voice frequency metallic (all others)	Any	25 dBrnC	36 dBrnC

[a] When a trunk comprises two facility types, the overall noise limit is the power sum of the individual limits for each facility type.

[b] A trunk should meet the *preservice limit* before being placed in service. A trunk found to exceed the *immediate action limit* should be removed from service or corrected immediately. Between these two limits, there exists a *maintenance limit* (not specified by Bellcore). Trunks exceeding the maintenance limit may be corrected on a scheduled basis.

[c] Tandem-connecting trunk with expected measured loss greater than 6 dB.

TABLE 3-4 STEADY NOISE OBJECTIVES: CARRIER SYSTEMS

From AT&T PUB 43801, *Digital Channel Bank Requirements and Objectives* (1982)[12]		
Condition	Idle channel noise (dBrnC0)	
	Requirement	Objective
End to end	23	20
Transmit section only	20	17
Receive section only	20	17

TABLE 3-5 STEADY NOISE OBJECTIVES:
SWITCHING OFFICES

From REA 522 (Digital Central Office)[14]	
Idle circuit	23 dBrnC0 maximum ≤18 dBrnC0 average
3-kHz flat	<35 dBrn0

From REA 524 (Analog Central Office)[15]	
Idle circuit	16 dBrnC maximum

Adapted from Bellcore *Notes on the BOC Intra-LATA Networks—1986*[13]				
	Noise limits (dBrnC)			
Office type	Steady state[a]		Average peak	
	Lower[b]	Upper	Lower	Upper
Analog	18	22	26	30
Digital	21	25	29	33

[a] Steady-state noise is the reading made by noting where the indication needle rests most of the time. Average peak noise is measured by setting the measuring set on damped response and mentally averaging the peak swings of the needle over a few minutes. (Bellcore does not provide guidance on measuring average peak noise using a set with a digital display.)

[b] If 20 through-office connections selected at random meet both the steady-state and average peak *lower* limits, the office is satisfactory. If four or more measurements (of either type) exceed the lower limit, or if a single measurement (of either type) exceeds the upper limit, the office is unsatisfactory. Offices with only a few measurements above the lower limit (and none above the upper limit) fall into a doubtful category. This situation is resolved by sampling more connections, as described in the Bellcore document.

TABLE 3-6 STEADY NOISE OBJECTIVES: SUBSCRIBER EQUIPMENT

From EIA-470 (Telephone Instruments)[5] and EIA-478 (Key Telephone Systems)[6]	
Condition	Maximum noise
On-hook, into 600 Ω	10 dBrnC −55 dBm (within the band 200–4000 Hz)
Off-hook, at 30–90 mA dc, into 900 Ω	15 dBrnC

From EIA-464 (PBXs)[4]	
Condition	Maximum noise[a] (on 95% of connections)
Off-hook	16 dBrnC 35 dBrn 3-kHz flat

[a] These limits were developed for analog PBXs. Their applicability to digital PBXs is under study.

TABLE 3-7 STEADY NOISE OBJECTIVES: TOTAL NETWORK[a]

Adapted from Bellcore *Notes on the BOC Intra-LATA Networks—1986*[13]			
Connection category[b]	Airline distance (mi)	Average noise (dBrnC)	Standard deviation (dB)
Short	0–180	19	6
Medium	180–720	23	5
Long	720–2900	27	4
Intercontinental	2900–12,500	30	3.5

[a] Subscriber-to-subscriber objectives.

[b] Intra-LATA connections are confined to the short, medium, and part of the long categories. However, the objectives shown were established predivestiture and Bellcore has provided the full long and intercontinental categories for completeness.

TABLE 3-8 IMPULSE NOISE OBJECTIVES: SUBSCRIBER LOOPS

Adapted from Bellcore *Notes on the BOC Intra-LATA Networks—1986*[13]	
Threshold	Limit
59 dBrnC	≤15 counts in 15 minutes on all loops measured at CO end

TABLE 3-9 IMPULSE NOISE OBJECTIVES: TRUNKS

Adapted from Bellcore *Notes on the BOC Intra-LATA Networks—1986*[13]		
Trunk type	Threshold (dBrnC0)	Limit[a]
Voice frequency	54	
Compandored[b] or mixed	66	
T-carrier[c]	62	≤5 counts in 5 minutes on 50% of trunks in a group; immediate action limit = 20 counts in 5 minutes
Noncompandored		
0–125 mi	58	
126–1000 mi	59	

[a] Limits still under study as of April 1986.

[b] Measure compandored circuits with a −13-dBm0 holding tone present.

[c] Holding tone not required when measuring T-carrier circuits.

TABLE 3-10 IMPULSE NOISE OBJECTIVES: CARRIER SYSTEMS

Adapted from *Digital Channel Bank Requirements and Objectives*[12] (PUB 43801)[a]	
Threshold (dBrnC0)	Maximum counts in 30 minutes
41	10
51	1
58	0.1

Adapted from Bellcore *Notes on the BOC Intra-LATA Networks—1986*[13]			
Carrier type	Condition	Threshold (dBrnC0)	Limits[b]
Compandored analog	−13 dBm0 holding tone	64	≤5 counts in 5 minutes on
Noncompandored	0–125 miles	56	50% of channels; imme-
	126–1000 miles	57	diate action limit = 20
T-carrier	No holding tone	60	counts in 5 minutes

[a] Reprinted with permission of AT&T. Copyright AT&T 1982; all rights reserved.

[b] Limits still under study as of April 1986.

TABLE 3-11 IMPULSE NOISE OBJECTIVES: SWITCHING EQUIPMENT

Adapted from Bellcore *Notes on the BOC Intra-LATA Networks—1986*[13]		
Office type	Threshold (dBrnC)	Limits
Crossbar Step by step Electronic	54 59 47	≤5 counts in 5 minutes on 50% of cross-office connections; immediate action limit = 20 counts in 5 minutes

From REA 522[14] and REA 524[15] (Digital and Analog Central Office)		
Office type	Threshold	Limits
Local digital Common control analog	54 dBrnC0 53 dBrnC	≤5 counts in 5 minutes on cross-office connections, measured 6 times during the busy hour

TABLE 3-12 IMPULSE NOISE OBJECTIVES: SUBSCRIBER EQUIPMENT

From EIA-464 (PBXs)[4]		
Threshold (dBrnC)		
Desired	Required	Limits
47	55	0 counts in 1 minute on 95% of connections

TABLE 3-13 IMPULSE NOISE OBJECTIVES: COMPLETE NETWORK

Adapted from Bellcore *Notes on the BOC Intra-LATA Networks—1986*[13]	
Threshold	Limits[a]
6 dB below received signal level	≤15 counts in 15 minutes on at least 85% of calls

[a] Limits still under study as of April 1986.

TABLE 3-14 CROSSTALK OBJECTIVES:
SUBSCRIBER LOOPS

From Bellcore *Notes on the BOC Intra-LATA Networks—1986*[13]	
	Crosstalk index[a]
	≤0.1% on 99% of loops

[a] The crosstalk index is the probability expressed in percent that one or more words from an interfering source will be intelligible in a call.

TABLE 3-15 CROSSTALK OBJECTIVES: TRUNKS

From Bellcore *Notes on the BOC Intra-LATA Networks—1986*[13]	
Trunk type	Crosstalk index[a] (%)
Interend office and tandem-connecting	≤0.5
Intertandem	≤1.0

[a] The crosstalk index is the probability expressed in percent that one or more words from an interfering source will be intelligible in a call.

TABLE 3-16 CROSSTALK OBJECTIVES:
CARRIER SYSTEMS

Adapted from *Digital Channel Bank Requirements and Objectives*[12] (PUB 43801)[a]		
Frequency (Hz)	Minimum equal-level crosstalk loss[b] (dB)	
	End to end	Near or far-end only
0–400	55	61
500	59	65
700–1100	65	71
3000	64	70
3400	59	65
3700–4000	55	61

[a] Reprinted with permission of AT&T. Copyright AT&T 1982; all rights reserved.

[b] Measured at 0 dBm0. Four-wire intrachannel crosstalk must also meet these requirements. This table is derived from a curve specified in PUB 43801.

TABLE 3-17   CROSSTALK OBJECTIVES: SWITCHING EQUIPMENT

From REA 522 (Digital Central Office)[14]		
Frequency	Condition	Minimum crosstalk loss
200–3400 Hz	0 dBm0	75 dB

From REA 524 (Analog Central Office)[15]		
Frequency	Condition	Minimum crosstalk loss[a]
1000 Hz	+10 dBm and 20 mA dc	96 dB (rms of 110 measurements) and 86 dB (negative 3 σ point)

[a] 100 random line-to-line plus 10 random line-to-trunk calls are made. Each crosstalk measurement is converted to a power. The powers are summed. The sum is divided by the number of readings (110). This results in the mean crosstalk power. When converted back to dB, the mean shall be $\geq$96 dB.

TABLE 3-18   CROSSTALK OBJECTIVES: STATION EQUIPMENT

From EIA-464 (PBXs)[4]			
		Minimum crosstalk loss	
Frequency	Condition	Desired	Required
200–3200 Hz	On 95% of connections	80 dB	75 dB
	On 100% of connections	—	70 dB

4 ≡ HYBRIDS

We begin this chapter with discussions of the two basic telephone transmission circuits: two-wire and four-wire. We then present the hybrid circuit, which converts between two- and four-wire circuits. Both passive and active hybrid implementations are described, and a detailed design example delves deeper into a hybrid circuit that uses an operational amplifier. Finally, transhybrid loss, a figure of merit for hybrids, is defined and its measurement demonstrated.

4-1 TRANSMISSION CIRCUITS

Two-Wire

The most familiar telephone circuit is the conventional *local loop* which runs from the central office to the subscriber's telephone set (Fig. 4-1a). This circuit uses a twisted pair of wires to help reduce hum and noise pickup.

Both directions of transmission share the pair. If you clip across the pair with a butt-in test set, you hear both sides of the conversation. Such a bidirectional circuit is called *two-wire* since it has been traditionally constructed with a pair of wires (literally, two wires).

Figure 4-1b shows that it is possible to construct a telephone circuit using a single wire plus a ground return. (Such a circuit is unbalanced with respect to ground and thus not very practical because of susceptibility to induced noise.)

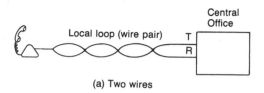

(a) Two wires

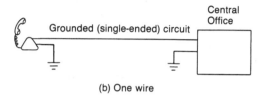

(b) One wire

FIGURE 4-1 Two-wire circuits.

Both directions of transmission appear on the single wire. For transmission purposes, we still refer to this as a two-wire circuit. (See Chapter 6 for a discussion of balanced and unbalanced circuits and their susceptibility to hum and noise interference.) Thus a *two-wire circuit* is defined as a single circuit (comprising one or two actual wires) that is shared by both directions of a telephone transmission.

Four-Wire

In Fig. 4-2 we show two telephone handsets connected by two pairs of wires. We have used a separate pair for each direction of transmission. The use of wire pairs reduces the susceptibility to induced noise. Such a circuit is called *four-wire* since it is traditionally constructed with two pairs, or four wires total.

It is also possible to construct a four-wire circuit using only two wires plus ground-return paths. We define a *four-wire circuit* as one with separate subcircuits for each direction of transmission. Each subcircuit may comprise one or two actual wires.

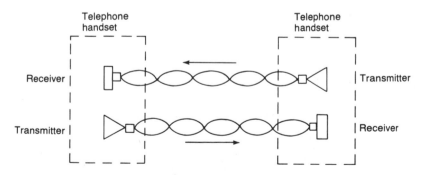

FIGURE 4-2 Four-wire circuit.

Be cautioned not to confuse the transmission concepts of two-wire and four-wire circuits with the implementation detail of how many physical wires are actually used. For example, a microwave link uses four-wire transmission even though it comprises no real wires.

Mixed Circuits

Figure 4-3 shows a simple two wire-to-four wire converter, also called a *Four-Wire Terminating Set* (4-WTS). Differential amplifier IC$_1$ and associated resistors perform the hybrid function (two wire-to-four wire conversion). We will return to this circuit later. For now, note the full complement of two-wire and four-wire circuits implemented with one, two, and four actual wires. Also note the use of some circuits that are not balanced with respect to ground. This is acceptable when signal lengths are short and appropriate design precautions are taken.

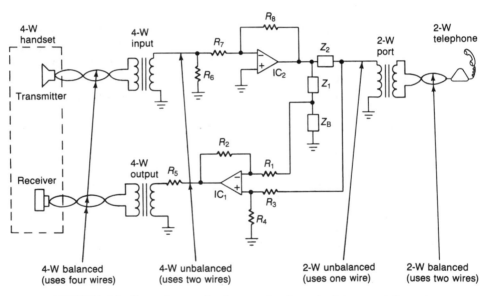

FIGURE 4-3 Four-wire terminating set showing mixed two- and four-wire circuits.

4-2 HYBRID CIRCUITS

Hybrid circuits (or simply *hybrids*) perform two wire-to-four wire conversion. These circuits are also known as *duplexers*. In Fig. 4-4 we have shown a hybrid as a black box. Signals from the four-wire port enter the hybrid on the four-wire input. These signals traverse path A and exit on the two-wire port. Signals from the two-wire port traverse path B and exit on the four-wire output. Signals enter-

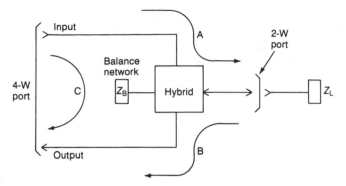

FIGURE 4-4   Hybrid circuit.

ing on the four-wire input should not appear on the four-wire output (i.e., signals should not traverse path C).

Hybrid Balance

A key element of the hybrid is the *balance network* (impedance Z_B). The value selected for Z_B is a function of Z_L, where Z_L is the line impedance connected to the two-wire port. When Z_B is correctly selected, the hybrid is said to be *balanced* and there is a minimum of signal transmission via path C.

Be aware that the term *balance* has at least two meanings in telephony. One meaning refers to circuits that have good longitudinal *balance*, that is, circuits that are well *balanced* with respect to ground (see Chapter 6). A second meaning, which we have just introduced, refers to the *balance* of a hybrid. This has to do with the matching of impedances and is similar in concept to a bridge circuit that is in *balance*. In this book we make our use of balance clear from the context.

Transhybrid Loss

Transhybrid loss (THL) is a figure of merit for a hybrid. The higher the THL, the better the balance and the less the coupling between input and output ports on the four-wire side. A high THL (low coupling) is desirable.

Transhybrid loss is found by applying a signal to the four-wire input port and measuring the signal resulting at the four-wire output port. The measurement is made at specified frequencies and with a specified Z_L connected to the two-wire port. Transhybrid loss is often measured with direct current flowing in the two-wire port. A measurement example is given later.

Hybrid Applications

Why do we want to convert telephone circuits back and forth between two- and four-wire? The standard telephone set has a two-wire interface so why not let all circuits be two-wire? The problem comes when we want to process the signals.

For example, we may want to amplify a two-wire signal that has been attenuated by a long cable. If we insert a simple unidirectional amplifier into the two-wire circuit, one direction of transmission will be amplified but the other direction will be blocked. One solution is to use hybrids to split the transmission into the two directions, then amplify each direction separately (Fig. 4-5a). The circuit within the dotted lines of Fig. 4-5a is called a *two-wire repeater*.

Another hybrid application allows modulation and demodulation in a radio link (Fig. 4-5b). A similar application uses hybrids to divide a two-wire circuit into transmit and receive directions as required for digital encoding and decoding in PCM carrier systems.

Within reach of you now there is probably a hybrid—the one inside your telephone set (Fig. 4-6). Here the hybrid splits the circuit into two directions for connection to the mouthpiece and earpiece.

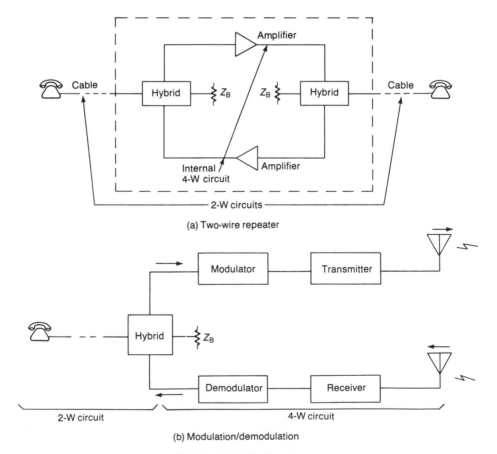

(a) Two-wire repeater

(b) Modulation/demodulation

FIGURE 4-5 Hybrid applications.

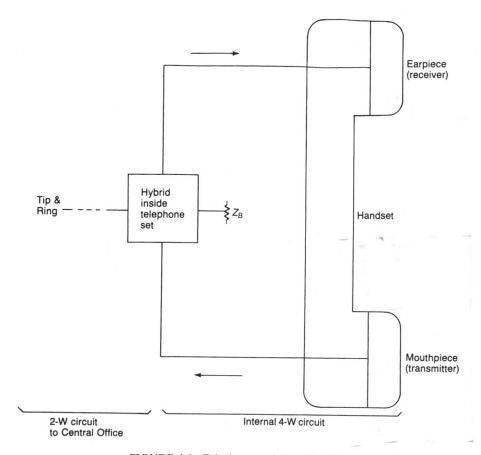

FIGURE 4-6 Telephone set internal hybrid.

At this point you may ask, "Why not make all circuits four-wire and dispense with hybrids?" That might make sense if we were starting from scratch, and indeed, the telephone network is evolving toward an all four-wire network. For the present, however, we have to contend with hundreds of millions of two-wire local loops and telephone sets. As the network evolved, it was cheaper to run two rather than four wires down the street and to provide two rather than four contacts on switching relays. Two-wire transmission and switching is still adequate in many case. Hybrids will see continued use in our current mixed two- and four-wire network.

Hybrid Implementations

Various electronic components are used to build hybrid circuits. We present three popular circuit configurations.

Two-transformer hybrid. Figure 4-7 shows a hybrid circuit implemented with two three-winding transformers. Note that:

1. T_1 and T_2 are identical and are assumed to be ideal.
2. Windings 3-4 and 5-6 have an equal number of turns.
3. The transformer windings are phased as shown.
4. $R_1 = R_2$.
5. For a balanced hybrid, $Z_B = Z_L$.

Circuit operation is as follows. Signals entering the input port cause voltages to appear at T_1 windings 3-4 and 5-6. Currents i_1 and i_2 then flow. Current i_2 develops a voltage across Z_L, thus delivering power out the two-wire port. Current i_1 causes power to be dissipated in Z_B.

If Z_B equals Z_L, then i_1 equals i_2. Due to the phasing of T_2's windings, i_1 and i_2 create equal but opposite flux in T_2. With no net flux in T_2, no power is delivered to the output port. That is, for a perfectly balanced hybrid, path C has infinite loss. The power from the input port splits evenly between Z_L and Z_B. Thus path A has a 3-dB loss (a 3-dB loss corresponds to half power; see Chapter 2).

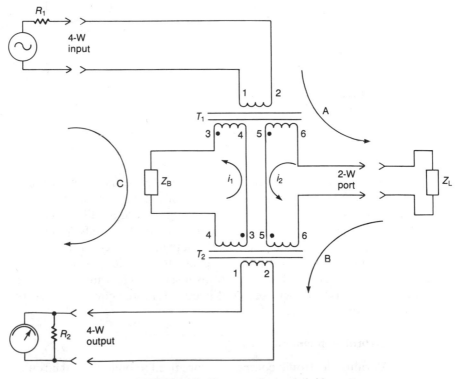

FIGURE 4-7 Two-transformer hybrid.

Signals entering the two-wire port create equal fluxes in the two transformers. Due to the winding phasing, i_1 equals 0 and no power is dissipated in Z_B. The signal from the two-wire port splits evenly between R_1 and R_2. Path B has a 3-dB loss. The impedance looking into the two-wire port is a function of the transformer turns ratios and R_1 and R_2.

Although basic, the circuit of Fig. 4-7 has considerable flexibility and several advantages:

1. All inputs and outputs are balanced with respect to ground.
2. Unequal impedances or unequal power splitting can be provided by adjusting the transformer turns ratios.
3. Any dc potentials present on a given port are isolated from the other ports by the transformers.
4. The circuit is passive.

One of the circuit's disadvantages is its 3-dB insertion loss.

Single transformer hybrid.   Figure 4-8 shows a hybrid circuit implemented with a single three-winding transformer. Note that:

1. For a balanced hybrid, $Z_L = 4Z_B$.
2. $R_1 = R_2$.
3. The transformer turns ratio and phasing are as shown.
4. We assume an ideal transformer.

Circuit operation is as follows. Signals entering on the input port are split on an equal power basis between Z_B and Z_L. When the hybrid is balanced, none of the input signal appears at the output.

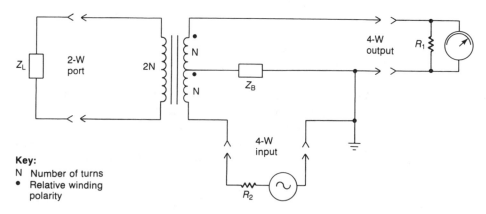

FIGURE 4-8   Single-transformer hybrid.

Since R_1 equals R_2, and since Z_B is located at the transformer's center tap, signals entering on the two-wire port are split equally between R_1 and R_2. Also, the net current through Z_B is zero.

The impedance seen looking into the two-wire port is R_1 plus R_2.

The single transformer hybrid has these advantages:

1. The two-wire port is balanced with respect to ground.
2. Any dc potential present on the two-wire port is isolated from the four-wire port.
3. The circuit is passive.

The circuit has these disadvantages:

1. The four-wire ports are not balanced with respect to ground.
2. There is a 3-dB insertion loss.

The general theory of transformer hybrids may be found in References 20 and 25. The two circuits presented above are but special cases of the general theory.

Op-amp hybrid.   The four-wire terminating set of Fig. 4-3 contains an *op-amp hybrid*. The hybrid alone is redrawn in Fig. 4-9. Z_L is the external impedance connected to the two-wire port. Impedances Z_1, Z_2, Z_B, and Z_L form a bridge. Op amp IC_1 and resistors R_1 to R_4 form a differential amplifier. This amplifier measures the voltage across two nodes of the bridge. The amplifier drives the four-wire output port. We will let R_1 to R_4 be much greater than the impedances in the legs of the bridge.

To analyze the four wire-to-two wire path, connect external circuit T to the two-wire port. Signals enter the input port from source V_T. Impedances Z_2 and Z_L drop this voltage to V_L at the two-wire port. Note that IC_1, Z_1, and Z_B do not effect V_L.

To analyze the two wire-to-four wire path, connect external circuit R to the two-wire port and assume that $V_T = 0$. Source V_L sees the impedance $Z_L + Z_2$. These impedances form a divider that drops V_L to a lower voltage at V_2.

Voltage V_1 is at ground potential (i.e., $V_1 = 0$). The values of Z_1 and Z_B do not effect V_2 or the impedance seen by V_L. Differential amplifier IC_1 amplifies $V_2 - V_1$ (which is simply V_2) and outputs the result at V_R, the four-wire output.

To analyze the transhybrid path, connect circuit Z to the two-wire port. Circuit Z has an impedance equal to that against which the hybrid is designed to balance. Impedance Z may or may not equal Z_L. Also, let $Z_1 = Z_2$ and assume that V_T is nonzero.

The transhybrid path is from the input port to the output port. Note that if $Z_B = Z$, the bridge is balanced; that is, $V_1 = V_2$ and $V_R = 0$. (To see this, recall that $Z_1 = Z_2$ and $Z_B = Z$. Now, the voltage drop across Z_1 equals the drop across Z_2,

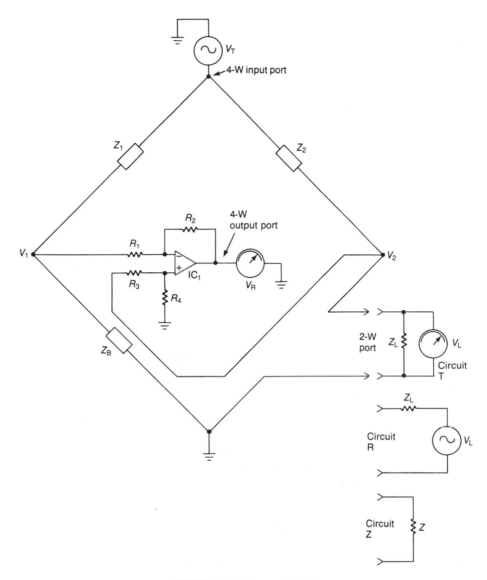

FIGURE 4-9   Op-amp hybrid.

and V_1 equals V_2.) Since V_R equals zero, no signal has reached the output port from the input port, and the hybrid is balanced.

Now if $Z \neq Z_B$ the bridge (and hybrid) are unbalanced, $V_1 \neq V_2$, and $V_R \neq 0$. The signal loss from V_T to V_R is the transhybrid loss. For a perfectly balanced hybrid, the THL would be infinite.

Note that for a balanced hybrid, Z_1 need not equal Z_2. If $Z_1/Z_B = Z_2/Z$, then there is still balance. This allows Z_1 and Z_B to be scaled up in impedance. This may be advantageous in some implementations, since it reduces the load on V_T.

The advantages of the op-amp hybrid are:

1. Amplifiers can compensate for the passive component losses.
2. The impedances are easy to adjust by changing the values of passive components.
3. The impedances can be scaled to provide practical component values.

The disadvantages are:

1. No ports are balanced with respect to ground.
2. There is no dc isolation between ports.

Note that in the following design example, the two disadvantages of the op-amp hybrid are overcome by using simple coupling transformers on all ports.

Example 4-1

We wish to specify the resistor values for the op-amp hybrid contained in the four-wire terminating set of Fig. 4-3. (Refer also to Fig. 4-9 for reference.)

Given: Ideal, 1:1 turns ratio transformers.
The four-wire external load and source impedances are 600 Ω.
The two-wire external load and source impedances are 900 Ω.
The hybrid is to balance against 900 Ω in series with 2.16 μF at the two-wire port.
A 2-dB power loss from port to port.
Impedance matching at all ports.
Ideal op amps.

Two-wire port   Make $Z_2 = 900$ Ω so that the two-wire port will see a 900-Ω input and output impedance as required for impedance matching.
Let $Z_1 = Z_2 = 900$ Ω.
Make $Z_B = 900$ Ω in series with 2.16 μF in order to balance the hybrid.
Let $R_1 = R_3 = 100$ kΩ so as not to load Z_1 and Z_2.

Input port   Make $R_6 = 600$ Ω so that the four-wire transmit port input will see 600 Ω as required for impedance matching.
Let $R_7 = 100$ kΩ so as not to load R_6.
R_8 will now determine the gain or loss in this direction; (see Fig. 4-10a). Let the input power = 0 dBm. This corresponds to an input voltage

$$V_1 = (1 \text{ mW} \times 600 \text{ } \Omega)^{1/2} = 0.775 \text{ V}$$

We want a 2-dB power loss through the 4-WTS, so the power out (measured by V_2) must equal -2 dBm. From the calculation method described in Chapter 2, a level

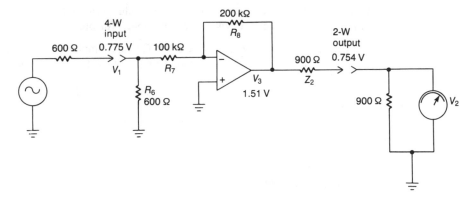

(a) Four to two-wire path

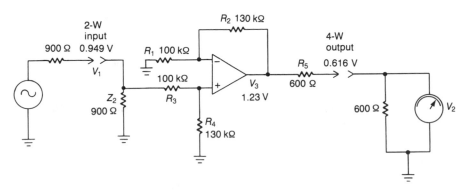

(b) Two to four-wire path

FIGURE 4-10   Gain/loss models: 4-WTS.

of -2 dBm into 900 Ω is

$$0.949 \text{ V} \times 10^{-2/20} = 0.754 \text{ V}$$

(We now have the four-wire input and the two-wire output both expressed conveniently in volts.)

The op amp's output

$$V_3 = \frac{Z_2 + 900 \ \Omega}{900 \ \Omega} \times V_2 = 1.51 \text{ V}$$

The amplifier voltage gain

$$\frac{V_3}{V_1} = \frac{1.51 \text{ V}}{0.775 \text{ V}} = \frac{R_8}{R_7} = 1.95$$

$$R_8 = 1.95 \times 100 \text{ k}\Omega = 195 \text{ k}\Omega$$

We will use 200 kΩ, the nearest standard 5% value. (Our actual calculated gain will then be -1.8 dB, which we will call close enough for our purposes.)

Output port   Make $R_5 = 600\ \Omega$ to provide a 600-Ω source impedance as required for impedance matching. Now, $R_2 = R_4$ and these resistors determine the gain in this direction (see Fig. 4-10b).

Let the input power = 0 dBm. This corresponds to an input voltage of $V_1 = (1\ \mathrm{mW} \times 900\ \Omega)^{1/2} = 0.949$ V.

For a -2-dB power gain, the output power equals -2 dBm. The corresponding output voltage

$$V_2 = 0.775\ \mathrm{V} \times 10^{-2/20} = 0.616\ \mathrm{V}$$

The op amp's output voltage

$$V_3 = \frac{R_5 + 600\ \Omega}{600\ \Omega} \times 0.616\ \mathrm{V} = 1.23\ \mathrm{V}$$

The amplifier's voltage gain equals

$$\frac{R_2}{R_1} = \frac{1.23\ \mathrm{V}}{0.949\ \mathrm{V}} = 1.30$$

$$R_2 = 100\ \mathrm{k}\Omega \times 1.30 = 130\ \mathrm{k}\Omega,\ \text{a standard 5\% value}$$

$$R_4 = R_2 = 130\ \mathrm{k}\Omega$$

We have now calculated values for all the components. Fig. 4-11 shows the completed 4-WTS.

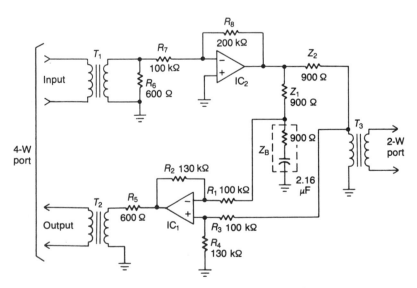

FIGURE 4-11   Complete four-wire terminating set.

Transhybrid Loss Measurement

Let's measure the THL of our completed 4-WTS (Fig. 4-12a). First we insert a known signal at the four-wire input. Then we measure the portion of this signal that reflects back from the two-wire port to the four-wire output. If the hybrid is balanced, there is no reflection.

Notice that any signal reflected from the two-wire port is attenuated 4 dB (2 dB in each direction). This 4 dB is an issue separate from the hybrid's degree of

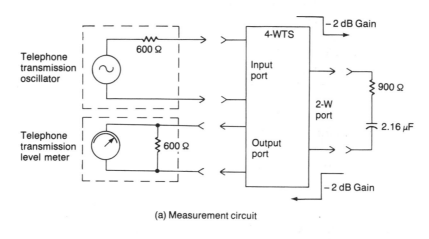

(a) Measurement circuit

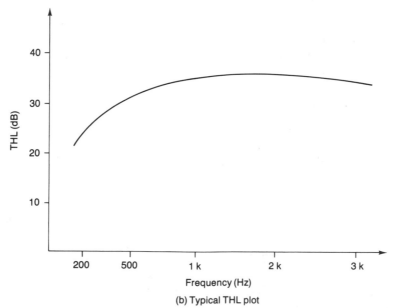

(b) Typical THL plot

FIGURE 4-12   Transhybrid loss.

balance. The 4-dB loss *does* appear in the raw measurement, however, so we must compensate for it.

Adjust the oscillator to deliver 0 dBm at 1 kHz into the 4-WTS input port. To check the setup, temporarily remove the 900 Ω in-series-with 2.16 μF external network. The level meter should read about −4 dBm. This indicates a complete reflection from the two-wire port (remember the 4-dB circuit loss). Now reconnect the network and note the meter reading. Let's say that it reads −42 dBm. This is a loss from the input port to the output port of 42 dB. Compensating for the built-in 4-dB loss, our hybrid has a THL at 1 kHz of 42 dB − 4 dB = 38 dB. This amount of unbalance is not unusual for a hybrid of moderate quality. In a real circuit it would be caused by component tolerances and the finite inductance of T_3.

We would next measure THL at other frequencies of interest and plot the results in Fig. 4-12b. It is typical for the THL of practical circuits to deteriorate at the edges of the voice band, as shown.

4-3 HYBRID STANDARDS AND SPECIFICATIONS

Hybrids are combined with other components (such as amplifiers, transformers, and other hybrids) to create communications circuits. Telephone industry standards usually specify the performance of complete circuits, not the performance of their individual electronic components. For this reason, we do not present here a list of industry requirements for hybrids. Hybrids do, however, influence circuit parameters that *are* specified. We show in Chapter 5, for example, how a hybrid's transhybrid loss can affect return loss in a built-up connection.

To give a feel for practical values, we list the THL specifications of hybrids used in three applications. Remember that these are manufacturer's specifications for typical products; these are not standards.

Test Instrument

The Northeast Electronics Model TN33 Test Hybrid is a self-contained, portable test instrument used for bench or field measurements. It contains a high quality two-transformer hybrid. Its inherent THL is specified as being better than:

60 dB: 200 to 500 Hz
65 dB: 500 to 2500 Hz
50 dB: 2500 to 10,000 Hz

Four-Wire Terminating Set

The Tellabs Model 4201 is a toll-grade 4-WTS constructed as a plug-in printed circuit board. It contains a two-transformer hybrid. Its THL is specified as better than 50 dB from 200 to 4000 Hz with up to 100 mA dc flowing in the two-wire port.

Private Branch Exchange

A typical Private Branch Exchange (PBX) may use a four-wire digital switching matrix to connect central office (CO) loops to PBX extensions. The CO interface circuits contain hybrids to connect the two-wire CO loop to the four-wire matrix. These may be op-amp hybrids, transformer-coupled at their two-wire port. Typically, the hybrid and transformer combination would be designed to provide 20 dB or better THL from 200 to 3400 Hz with 20 to 70 mA dc flowing in the transformer.

As shown above, THL performance can vary from 20 dB to 65 dB as we go from customer equipment to instrumentation-grade products.

5 RETURN LOSS and ECHO

Echo is a phenomenon well known to telephone users. On a long-distance call you may hear an echo of your own voice or an echo of the distant party's voice. The same circuit conditions that cause echo on long-distance calls can cause a hollow sound on short-distance calls.

In this chapter we explain why echo occurs in the network and we define return loss—a quantity that is measured to assess echo performance. Return loss is further classified by type, such as echo return loss and singing return loss. Related topics include echo cancelers, singing, hybrid balance networks, and office balance. We also present return-loss instruments, measurement examples, industry objectives, and surveyed performance.

5-1 ECHO AND REFLECTIONS

5-1-1 Echo

Echo is signal power that is reflected back toward the signal's origin. In two-wire circuits, a reflection occurs at a point of impedance mismatch. Reflections also occur at two wire-to-four wire junctions. If a circuit is four-wire end to end with four-wire terminations, it has no voice-frequency reflections. Although a built-up connection may comprise many reflection points, the echo from one reflection usually predominates.

5-1-2 Reflections

Figure 5-1 shows sources of reflections in the telephone network. In the connection of Fig. 5-1a, the cable plant contributes impedance discontinuities at gauge changes, load coils, and bridged tap. (A *bridged tap* is a short branch off the main cable. Electrically, a pair in the branch cable connects in parallel to a pair in the main cable.) There are also reflections from the impedance mismatches between the cable and the end office and between the cable and the far-end telephone.

When loop electronics are involved, there are reflections from repeaters and station carrier systems (Fig. 5-1b). Manufacturers and users attempt to match the impedance of loop electronics to various cable schemes. In practice, the degree of match is quite variable, due to the nonuniformity of subscriber cable layout.

End offices (Fig. 5-1c), tandem offices, and trunks of all kinds (Fig. 5-1d) also contribute reflections. We will return to Fig. 5-1 as we examine specific echo situations.

5-1-3 Echo Paths

Figure 5-2 shows a basic circuit connecting two telephones. The circuit consists of two two-wire portions at each end plus a four-wire portion in the middle. Hybrids are used to convert between two- and four-wire circuits. Amplifiers in the four-wire portion recover the signal power that is lost in the hybrids. The amplifiers may also provide gain to compensate for loss in the two- and four-wire portions.

The primary speech path (Fig. 5-2a) extends from the west telephone to the east telephone. (We have arbitrarily assigned west as the talker.)

Talker echo.   If part of the signal is reflected from the east hybrid back to the west telephone, *talker echo* is present (Fig. 5-2b). Such a reflection occurs if the east hybrid balance network does not match the impedance of the associated two-wire circuit. That is, if the east hybrid has finite transhybrid loss. [Transhybrid Loss (THL) was defined in Chapter 4.]

All paths in the network have some delay. The talker echo path has twice the delay of the primary speech path. On a long connection with echo, the talker can hear a delayed version of his or her own voice. This can be quite distracting. If the echo delay *and* the echo power are both great enough, the talker's speech process may be impaired. (This phenomenon can be demonstrated by recording your voice on a tape recorder having separate record and play heads. While recording, listen with headphones to the delayed monitor signal from the play head.)

On short connections, talker echo will have less delay and will simply sound like louder sidetone. (*Sidetone* is the sound of your own voice heard in a telephone receiver when you talk into the transmitter.) Some sidetone is desired (so the phone will not sound dead), but too much sidetone can cause the talker to speak more softly. This may prevent the distant party from hearing the talker well.

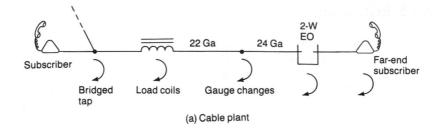

(a) Cable plant

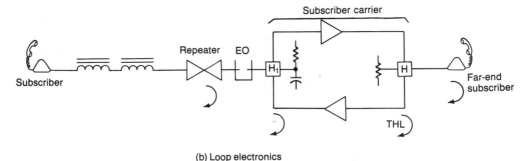

(b) Loop electronics

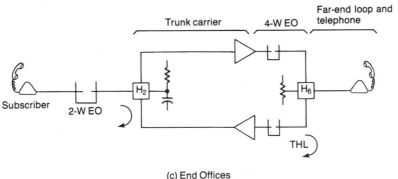

(c) End Offices

FIGURE 5-1 Source of reflections in the telephone network.

Listener echo. *Listener echo* occurs when the speech power reflects a second time—this time from the hybrid near the talker. The listener echo path is shown in Fig. 5-2c. When this impairment is present, the listener hears the talker via both the primary speech path and the listener echo path. On long connections, there will be a delay between these two paths that causes the listener to hear an echo of the talker's voice. This can be objectionable to voice users and can cause errors in voice-band data.

On short connections, with little delay, the talker's voice may sound hollow to the listener (as if the talker were in a barrel). The primary speech signal and the

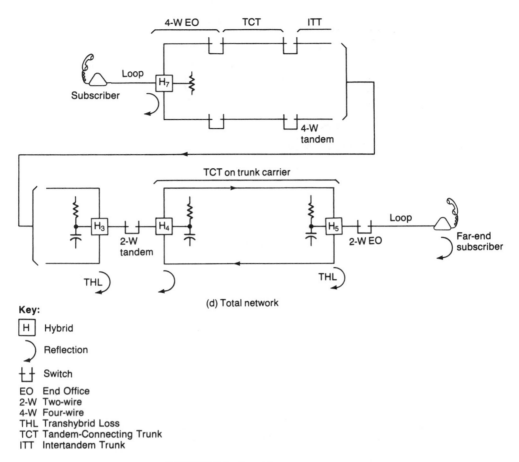

Key:

H	Hybrid
	Reflection
	Switch
EO	End Office
2-W	Two-wire
4-W	Four-wire
THL	Transhybrid Loss
TCT	Tandem-Connecting Trunk
ITT	Intertandem Trunk

(d) Total network

FIGURE 5-1 (*cont.*)

listener echo signal add and subtract at various frequencies, creating peaks and dips in the frequency response. This contributes to the distracting sound of such calls.

Singing.   In Fig. 5-2c, note that the signal could continue to reflect back and forth in the four-wire portion of the connection (between the hybrids). The net gain and phase at a particular frequency may be such that the circuit can sustain oscillation at that frequency. This is called *singing*. Real circuits tend to sing at frequencies near the low or high ends of the voice band. This is because impedance mismatch is more pronounced at the edges of the band. Singing sounds like a loud tone and renders a circuit unusable. Since it usually occurs at high levels, singing may also crosstalk into adjacent circuits.

Singing may be thought of as listener echo that is so great that the circuit oscillates. Conversely, circuits with listener echo but not quite enough gain to

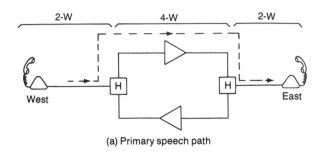

(a) Primary speech path

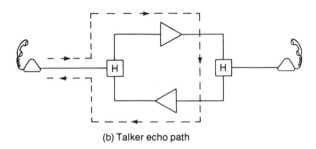

(b) Talker echo path

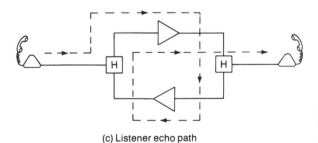

(c) Listener echo path

FIGURE 5-2 Echo paths. (Adapted from Bellcore *Notes on the BOC Intra-LATA Networks—1986.*[13])

sustain oscillation are said to exhibit *near-singing*. *Near-singing* and *unstable* are both terms that describe the hollow sound of listener echo on short connections.

Multiple Echoes

The circuit in Fig. 5-2 is a simplification that is useful while defining terms. In the more general case, the two-wire portions of this circuit may comprise two-wire trunks joined at two-wire switches. Similarly, the four-wire portion may contain a mixture of four-wire trunks and switches. Furthermore, a real built-up connection is likely to contain several two- to four-wire conversions. This all leads to a multitude of echoes that contribute to the total talker and listener echo levels.

The design of the telephone network is such that on most connections, one echo source predominates—the echo from the far end.

Total network echo.   Figure 5-3 shows a simplified view of a long connection through the network. The impedance mismatches at switching offices are minimized for calls passing *through* the offices. For calls that terminate in an office, there may be a significant echo from the junction between the subscriber loop and the rest of the network.

Returning to Fig. 5-1, we can see in more detail two examples of the predominate echo from the far end. Figure 5-1c shows a distant four-wire end office. The level of talker echo is a function of how well the hybrid in the distant end office matches the far-end loop and telephone set. Another example (Fig. 5-1d) involves a distant two-wire end office reached by a trunk carrier system. Here the level of talker echo is a function of how well the distant hybrid in the carrier system matches the far-end loop and telephone set. (The two-wire end office will appear fairly transparent in this example.) The matching of impedances at switching offices is called *office balance* and is discussed later.

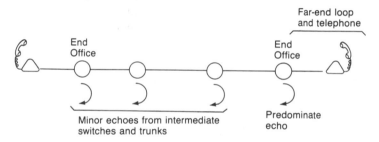

FIGURE 5-3   Echo in the total network.

Control of Talker Echo

Matching impedances.   Recall that the subjective annoyance of talker echo is a function of echo level *and* echo delay. The first step in reducing echo level is to assure minimal reflections from intermediate offices. This is done by controlling the *through balance* and *terminal balance* of tandem switches as described later. Next, the end office reflection is controlled by balancing the hybrid nearest the subscriber loop. This balance is only a compromise in most end offices, due to the variety in subscriber loops. It does help, however, if the loops are designed with an attempt made to achieve uniform impedance.

Inserting loss.   In addition to impedance matching, echo control uses intentionally inserted loss to attenuate the echo. Note that each decibel of loss inserted symmetrically into the primary speech path results in a 2-dB attenuation of talker echo. (*Symmetrical loss* is equal loss in each direction of transmission and is usually the rule.) Since some loss can be tolerated in the primary path, this scheme works well on short and medium-distance connections. As an example, the loss plan for digital networks allows 6 dB of loss between end offices. This provides 12 dB of talker echo reduction.

Using echo suppressors and cancelers.   As the connection distance and thus echo delay increase, the subjective echo annoyance also increases. To compensate, more loss could be added to the circuit. However, there comes a point where the primary path loss would be too great for the users to communicate comfortably. For purposes of setting design rules, this point has been defined as 1850 miles. Beyond this distance, network design rules call for an *echo control device* to be inserted.

Echo suppressors are older echo control devices. An echo suppressor is installed in a four-wire trunk, where it monitors the levels on each direction of transmission. The suppressor decides which direction is being used for the primary speech path and turns off the transmission in the other direction. This breaks the echo path. When one party stops talking and the other starts, the suppressor detects the shift in levels and reverses the circuit.

There may be some speech clipping when an echo suppressor switches directions. If you want to interrupt the talker, you may have difficulty capturing the circuit. These devices essentially convert a full-duplex circuit into a half-duplex circuit with automatic direction control. Echo suppressors disable themselves when they detect *data conditioning tone* from modems. This allows modems to use both directions of transmission simultaneously.

Echo cancelers are the currently preferred echo control device. Echo cancelers introduce fewer impairments into a circuit than do echo suppressors. An echo canceler is installed in the transmit path at each end of the four-wire portion of a built-up connection (Fig. 5-4).

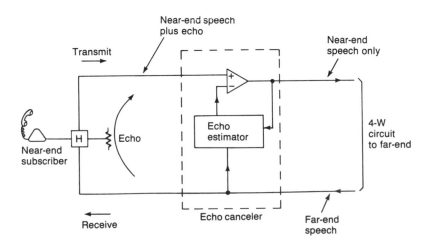

FIGURE 5-4   Echo canceler.

In operation, a part of the canceler called an *echo estimator* uses digital signal processing to estimate the echo path characteristics. (Here, the echo path starts at the point where the estimator taps the receive leg, extends through the hybrid, and ends back at the echo canceler. The four-wire circuit between the echo canceler and the hybrid is called the *tail*.) The far-end speech is processed in the canceler to create a signal approximately equal to the echo. This estimated echo is *subtracted* from the transmit signal so that only near-end speech returns to the far end. Another echo canceler at the far end blocks echoes at the far-end hybrid. This is known as split operation, and the canceler in Fig. 5-4 is called a *split echo canceler*.

The echo estimator continually updates its characterization of the echo path. However, when both parties speak (*double-talking*), the updates stop. During double-talking, echo cancellation continues and a full-duplex circuit is maintained. This is a great operational advantage over echo suppressors.

Control of Listener Echo

For long connections, the steps taken to control talker echo will also control listener echo. For short connections, listener echo in the form of near-singing and singing is a potential problem. In many of these applications it is not acceptable to insert extra loss in an attempt to attenuate echo. Instead, singing is controlled by more precise hybrid balancing. Examples are given later in this chapter when we discuss hybrids.

5-2 RETURN LOSS

A signal traveling down a transmission path can be reflected from an impedance discontinuity and return to the near end. The level of the return signal relative to the transmitted signal is the *return loss*. Return loss is a function of frequency, and there are various ways to measure return loss that take this frequency dependence into account. In the following sections we define return loss in several ways.

5-2-1 Return Loss Defined by Circuit Configuration

Four-Wire Return Loss

Four-wire return loss is measured at a four-wire point and is defined (in dB) as

$$10 \log \frac{\text{transmitted power}}{\text{received power}} \tag{5-1}$$

Figure 5-5a shows the basic four-wire return loss test setup.

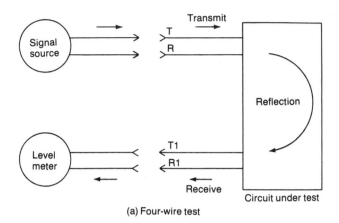

(a) Four-wire test

(b) Two-wire test

FIGURE 5-5   Basic return loss tests.

Two-Wire Return Loss

Two-wire return loss is measured at a two-wire point and requires a *test hybrid* as part of the measuring equipment (Fig. 5-5b). (A test hybrid is a precision hybrid designed with a variety of selectable impedances and calibration capabilities.) Two-wire return loss (in dB) is defined as

$$10 \log \frac{\text{transmitted power}}{\text{reflected power}} \qquad (5\text{-}2)$$

The reflected power is a function of the transmitted power *and* of the degree of impedance mismatch at the pont of reflection. In terms of circuit characteristics, two-wire return loss (in dB) is defined as

$$20 \log \left| \frac{Z + Z_{\text{ref}}}{Z - Z_{\text{ref}}} \right| \qquad (5\text{-}3)$$

where Z is the impedance of the circuit on the far side of the discontinuity, and Z_{ref} is the impedance on the near side. When Eq. (5-3) is applied to Fig. 5-5b, Z is the impedance of the circuit under test, and Z_{ref} is the balance network impedance.

Reflection coefficient. In some fields of communications, *reflection coefficient* is used to describe the same impairment that we call return loss in telephony. By definition,

$$\rho = \frac{Z - Z_{\text{ref}}}{Z + Z_{\text{ref}}} \qquad (5\text{-}4)$$

where ρ is the reflection coefficient and Z and Z_{ref} are as defined above. The ratio ρ has no units. Return loss and reflection coefficient are related by the formula

$$\text{return loss (in dB)} = -20 \log |\rho| \qquad (5\text{-}5)$$

Example 5-1

A transmission line with a 600-Ω characteristic impedance is terminated by a device with an impedance of 700 Ω. What are the return loss and reflection coefficient at the termination?

First we find the return loss using Eq. (5-3):

$$\text{return loss} = 20 \log \left| \frac{700 \ \Omega + 600 \ \Omega}{700 \ \Omega - 600 \ \Omega} \right|$$

$$= 22.3 \text{ dB}$$

Next we find the reflection coefficient using Eq. (5-4):

$$\rho = \frac{700 \ \Omega - 600 \ \Omega}{700 \ \Omega + 600 \ \Omega}$$

$$= 0.077$$

5-2-2 Return Loss Defined by Frequency Characteristics

Single-Frequency Return Loss

Single-frequency return loss is simply return loss measured using a single-frequency sine wave. Single-Frequency (SF) return loss may also be specified over a range of frequencies: for example, "minimum return loss at any frequency in the range 200 to 500 Hz shall be 16 dB." When measuring SF return loss, individual readings are taken at discrete frequencies. This is in contrast to weighted return loss (defined later), which is measured simultaneously over all frequencies in a band. Of the two measurements, SF return loss is more diagnostic since it can show the exact frequency at which the worst return loss occurs.

Impedance match. In voice telephony, equipment impedance is often specified as a nominal value with no tolerance given. Elsewhere in the specification, the required return loss may be stated. These two specifications are used together to find the tolerance allowed on the equipment's impedance. That is, the equipment's impedance tolerance is specified in terms of return loss, not percent error.

Example 5-2

Assume that the specification for a PBX trunk circuit requires a nominal impedance of 600 Ω. The specification also requires a minimum return loss of 15 dB measured with a reference impedance equal to the nominal impedance. For what input impedance should you design the trunk circuit?

The design should be 600 Ω nominal. The real question here is how much deviation from 600 Ω is allowed? We must start with the return loss and calculate the impedance. Substitution into Eq. (5-3) yields

$$15 \text{ dB} = 20 \log \left| \frac{Z + 600 \ \Omega}{Z - 600 \ \Omega} \right|$$

where Z is the trunk circuit's input impedance. There are two solutions: $Z = 419 \ \Omega$ and $Z' = 860 \ \Omega$. At these impedances, the return loss is exactly 15 dB. To meet the return loss specification, the trunk circuit's impedance must be between 419 and 860 Ω. The closer the impedance is to 600 Ω, the better the return loss.

Weighted Return Loss

Weighted return loss is measured over a band of frequencies. A number of frequency bands are used, each for a specific purpose, as explained shortly. Measuring weighted return loss is not as diagnostic as measuring SF return loss. However, weighted return loss can be measured more quickly and serves as a figure of merit. If weighted return loss is poor, SF return loss can also be measured in an attempt to diagnose the problem. Weighted return loss is measured using white noise that is filtered with the appropriate weighting curve. The filters can be located in either the sending or receiving part of the measurement circuit.

Echo Return Loss. Echo Return Loss (ERL) is weighted return loss measured in the middle of the voice-frequency band. In its Standard 743-1984,[8] the Institute of Electrical and Electronics Engineers (IEEE) specifies the ERL weighting characteristic in tabular form. We have plotted this weighting curve in Fig. 5-6. The ERL weighting curve is a combination of three frequency characteristics:

1. Energy content versus frequency of the human voice
2. Frequency response of a 500-type telephone set
3. Sensitivity versus frequency of the human ear

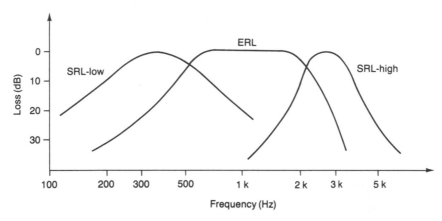

FIGURE 5-6   Return loss weighting characteristics. (Adapted from ANSI/IEEE Std 743-1984, *IEEE Standard Methods and Equipment for Measuring the Transmission Characteristics of Analog Voice Frequency Circuits.*[8])

Thus echo return loss measures the subjective effect of echo on human beings using standard telephones.

Singing Return Loss.   *Singing Return Loss* (SRL) is weighted return loss measured in two bands, one at the extreme low end of the voice band and one at the extreme high end. These two measurements are known as *SRL-low* and *SRL-high*. The IEEE[8] specifies the SRL weighting characteristics, and we plot them in Fig. 5-6. If singing or near-singing occur, they are likely to do so at the extremes of the voice-frequency range. Thus SRL is a measure of a circuit's potential for singing and instability. SRL-low and SRL-high are measured separately. Singing return loss is then stated as the lowest (worst) of the two readings.

5-2-3 Miscellaneous Definitions

Singing Point

Singing point measurement predates and has been replaced by singing return loss measurement. Nevertheless, references to singing point are still seen. To measure singing point, gain is inserted into the four-wire portion of the circuit under test. The gain is increased until the circuit just starts to sing; then the frequency at which the circuit sings is noted (this frequency usually falls within one of the SRL bands). The extra inserted gain is removed, then single-frequency return loss is measured at the previously noted singing frequency. The result of this SF return loss measurement is the singing point.

An alternative definition of singing point states that the singing point equals the gain added to produce singing. In this definition there is no separate SF return loss measurement. In either case, the singing *frequency* should be reported as part of the singing point (e.g., "15 dB at 220 Hz"). We recommend using the newer and simpler SRL measurement, which yields the same result as a singing point measurement.

Singing Margin

Singing margin is the amount of gain that must be added to the four-wire portion of a circuit to just start it singing. Singing margin is more useful as a concept than as a quantity to be measured.

Structural Return Loss

Structural return loss is measured in the same way as single-frequency return loss. However, the measurements are made over the full voice band, and the results are plotted as return loss versus frequency. These plots are often made on newly installed cable pairs as part of a cable acceptance test. Shorted load coils and other construction defects that might be missed using weighted return loss measurement show up easily on structural return loss plots. When structural return loss is measured, precision balance networks (described later) are used as both the reference impedance and as the far-end terminating impedance.

The terms *singing return loss* and *structural return loss* are sometimes confused because of their initials. The two measurements are not the same.

On-Hook Return Loss

The discussion so far has assumed that return loss is measured in the talking or *off-hook* state of the circuits involved. *On-hook return loss* is also important in situations where circuit stability depends on good impedance match regardless of the circuit's signaling state. This means that equipment such as PBX trunk circuits may have to be designed for a minimum on-hook singing return loss if good *on-hook stability* is to be achieved.

5-3 RETURN LOSS MEASUREMENT

In this section we present information about instruments that measure return loss in all its various forms. You will also find examples showing a variety of test situations.

5-3-1 Return Loss Instruments

Figure 5-7 shows an instrument capable of measuring combinations of two- or four-wire, single-frequency or weighted, and on- or off-hook return loss. This instrument requires considerable manual setup and control; however, it provides

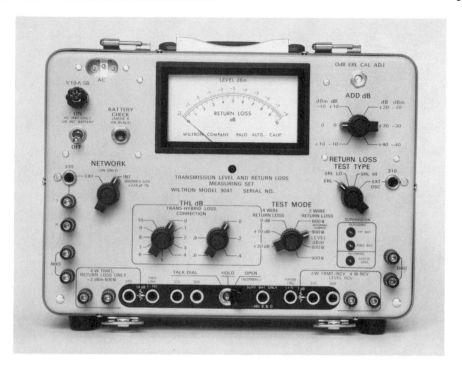

FIGURE 5-7   Return loss measuring set. (Courtesy of Wiltron Company.)

great measurement flexibility, including provisions for two-wire battery feed and external balance networks.

Some of the newer, multipurpose test instruments claim return loss measurement among their capabilities. Be cautioned that the measurements offered may be only a subset of the full return loss repertoire. For example, some instruments contain no test hybrid, so are limited to measuring four-wire return loss. Other instrument features to look for include ERL and SRL weighting filters, and terminals for connecting an external balance network. The Design Development Inc. Model 100P-4 is an example of a multipurpose instrument that includes all the above mentioned return loss features and capabilities except battery feed. (Battery feed can be provided externally as shown in Fig. 5-9.)

Functional Description

The IEEE Standard 743-1984[8] specifies features and performance for return loss measuring sets. Figure 5-8 is a block diagram of a return loss set based on the IEEE standard and typical commercial designs. The instrument can be configured for testing either two-wire or four-wire return loss.

Two-wire test. In Fig. 5-8a, the test signal originates from a white noise generator and is shaped into the ERL, SRL-low, and SRL-high spectra by three

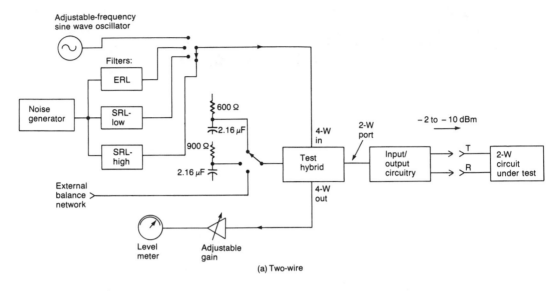

(a) Two-wire

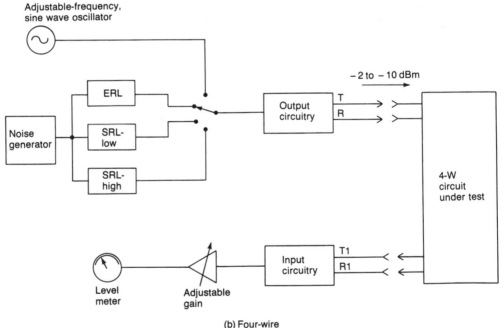

(b) Four-wire

FIGURE 5-8 Return loss measuring set: functional diagrams. (Diagrams adapted courtesy of Wiltron Company.)

filters. A fourth signal source is provided by a variable-frequency sine-wave oscillator. The test signal source (single-frequency or weighted in one of the three bands) is switched to the four-wire input of the test hybrid. The signal exits the hybrid's two-wire port and passes through the output circuitry to the circuit under test. The output circuitry provides 600- or 900-Ω impedance matching, dc hold coil (switchable in or out), and optional dc battery feed.

The test signal reflects from the circuit under test, returns to the test hybrid, and exits from the hybrid's four-wire output port. Adjustable gain is provided at this point to compensate for test hybrid and other test set losses. Finally, the return loss is read directly in decibels on the level meter. The required accuracy is ±0.5 dB or better over a range of 0 to 40 dB return loss.

The balance network for the test hybrid is selectable. Two internal networks (600 and 900 Ω each in series with 2.16 μF) and access to an external network are provided.

Four-wire test.   The instrument's four-wire configuration (Fig. 5-8b) is the same as its two-wire configuration except as follows. There is no test hybrid or balance network in the four-wire mode. The interface to the circuit under test is four-wire, requiring separate input and output circuitry. In the four-wire mode, the minimum measurement range is 0 to 50 dB return loss. Although IEEE Standard 743-1984 recommends an instrument output level of −10 dBm, this is too high for use with four-wire −16-TLP carrier channels. In this application, additional attenuation is required either internally or externally to the test set.

Test circuit alternatives.   The IEEE allows some variation in the test circuit. The weighting function may be in the receive leg instead of the transmit leg, for example. Also, weighted return loss may be replaced by a stepped or swept frequency measurement. In either of these cases, the individual step or swept return loss measurements must be averaged on a power basis.

5-3-2 Measurement Examples

Example 5-3

We wish to measure the return loss of a telephone set for compliance with industry standards. The Electronics Industries Association (EIA) specifies the SF return loss of telephone sets to be ≥3.5 dB (200 to 3200 Hz) and ≥7 dB (500 to 2500 Hz) when measured against 600 Ω (Table 5-5).

The measurement must be made in the off-hook state with dc flowing equal to that supplied by a loop simulator circuit. (The EIA loop simulator was presented in Chapter 1.) Assume we have found that the loop simulator delivers a range of 22 to 90 mA dc to the telephone.

Equipment   A two-wire single-frequency measurement is called for (Fig. 5-9). Since the reference impedance of 600 Ω is not available internally, we use an external balance network. (Although 600 Ω *in series with 2.16 μF* is a standard built-in network, 600 Ω with no capacitor is not.) Single-frequency return loss calls for use of a sine-wave oscillator. We show an oscillator external to the measuring set, but many sets have one built in.

In this example we are using a return loss set that has a provision for supplying dc current to the equipment under test. If the set does not have this provision, an

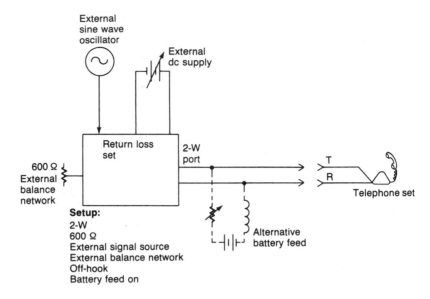

FIGURE 5-9   Measuring single-frequency return loss of a telephone set.

external battery feed circuit (shown dotted in Fig. 5-9) is required. Even if the set does have internal battery feed capability, the internal feed circuit may saturate at high currents. This would also be reason to use an external feed circuit.

Calibration   Calibration is verified by opening or shorting the instrument's two-wire port and checking for a 0-dB return loss. (Since there is complete signal reflection from an open or short, a 0-dB return loss reading should result.) The level of the external oscillator and possibly other instrument controls should be adjusted as necessary to achieve a 0 ± 0.5 dB calibration reading.

Data   Let's assume that the measurements yield the data shown in Table 5-1. Examining the table we find that over the full range 200 to 3200 Hz, the return loss stays

TABLE 5-1   DATA FOR EXAMPLE 5-3

Frequency (Hz)	Return loss (dB)	
	At 22 mA dc	At 90 mA dc
200	5.0	4.5
300	5.5	5.0
500	7.0	6.0
1000	12.5	12.5
2000	13.0	13.0
2500	12.0	12.0
3000	12.0	12.0
3200	11.0	11.0

above the specified minimum of 3.5 dB. However, at the higher dc current and at 500 Hz, the equipment is 1 dB "out of spec" (compare the 7-dB specification with the 6-dB measurement). This defect could be caused by a transformer that loses inductance at high dc currents and thus has a decreased impedance at low frequencies.

Example 5-4

We wish to measure the *weighted transhybrid loss* of a two-wire channel unit in a digital carrier system. [Transhybrid Loss (THL) was covered in Chapter 4. Weighted THL is simply THL measured using an ERL or SRL noise spectrum.] AT&T PUB 43801,[12] the applicable document, specifies the THL to be ≥34 dB with echo weighting and ≥20 dB with singing weighting. AT&T also specifies a test circuit and procedure that we approximate in this example.

Equipment   Figure 5-10 shows the test setup. The Channel Unit (CU) under test is terminated in the impedance against which its hybrid was designed to balance. We have picked a 900-Ω CU which is designed to balance against 900 Ω in series with 2.16 μF. This CU does not use the tip and ring leads for signaling, so we need not worry about providing a direct current in these leads. We will be measuring the THL of this far-end CU.

To establish a connection to the four-wire ports of the hybrid in the CU under test, we set up a digital connection to a four-wire CU. This provides four-wire access to the hybrid under test. *A four-wire return loss* measurement (described above) is now made at the four-wire CU.

Calibration   Between the return loss set and the far-end hybrid, there are gains and losses that are part of the carrier system design. During calibration, adjustments are made to compensate for these gains and losses. The setup is calibrated by removing the terminating network from the CU under test. This creates a full reflection and a reference return loss. The return loss set's receive calibration control (sometimes called *THL correction*) can now be adjusted for an indication of 0 dB return loss.

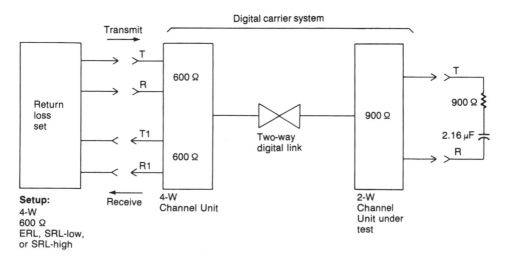

FIGURE 5-10   Measuring weighted transhybrid loss of a two-wire channel unit.

When the far-end network is restored, the THL of the CU under test is read directly on the return loss set.

Data Measurements are made using three weightings: ERL, SRL-low, and SRL-high. If the ERL reading is ≥34 dB and both SRL-low and SRL-high are ≥20 dB, the CU meets the specification.

Example 5-5

In this example we measure the impedance of the individual four-wire ports of a digital channel unit. AT&T PUB 43801 specifies that these ports have an impedance of 600 Ω with an allowable deviation from 600 Ω stated in terms of return loss. At 1000 Hz this return loss should be ≥28 dB, and over the range 300 to 3000 Hz it should be ≥23 dB. The verification of this specification calls for a two-wire single-frequency return loss measurement (Fig. 5-11).

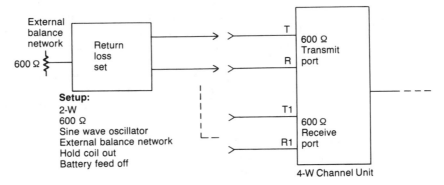

FIGURE 5-11 Measuring return loss of individual four-wire ports.

Equipment Do not allow a *two-wire* measurement on a *four-wire* CU be a source of confusion. As Fig. 5-11 shows, the four-wire ports are being measured one at a time. This calls for a two-wire test configuration.

In this example, our return loss set has a built-in sinewave oscillator, but we do need an external balance network. Other test set control settings are shown in Fig. 5-11.

Data Return loss data are taken at frequencies from 300 to 3000 Hz (including a data point at 1000 Hz), then compared with the specification. Next the set is connected to the CU receive port and the measurements are repeated. When measuring the receive port, either idle code or an analog quiet termination should be applied toward the CU under test from the far end.

5-4 HYBRIDS

Using our new knowledge of return loss and echo, we resume the discussion of hybrids as promised in the preceding chapter.

5-4-1 Return Loss and Transhybrid Loss

Consider the trunk carrier link shown in Fig. 5-1d. When the subscriber's speech reaches the west end of this link, there will be a reflection from hybrid H_4. This reflection is measured as the return loss at the two-wire port of the west channel unit.

The same carrier system provides another reflection at hybrid H_5 in the east CU. This reflection is measured as the transhybrid loss of H_5.

It may be useful to combine these two reflections into one quantity that covers the end-to-end performance of the trunk carrier. The following example shows how.

Example 5-6

The carrier link of Fig. 5-1d is shown in more detail in Fig. 5-12a. We have assumed the use of digital carrier that conforms to AT&T PUB 43801.[12] Since this is a tandem-connecting trunk, we have assigned an Inserted Connection Loss (ICL) of 3 dB. This loss is implemented using analog pads at the receive end of each direction. We have assumed that the west switching office has a 600-Ω nominal impedance and that the east switch is a 900-Ω office. Since we deal with *power* levels, this difference does

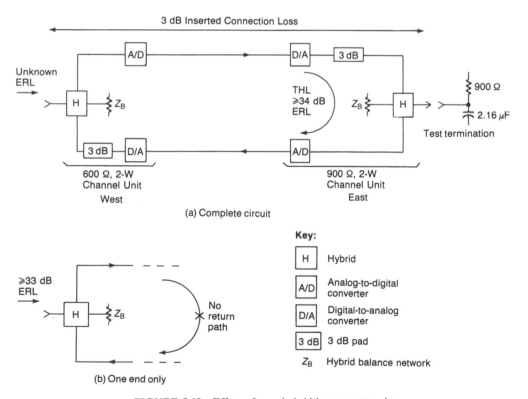

FIGURE 5-12 Effect of transhybrid loss on return loss.

not affect our example, but it does illustrate the variety of conditions to be found in practical cases.

East   AT&T specifies separately the reflections at each end of this link. The trans-hybrid loss of the east hybrid is specified as ≥ 34 dB ERL. (This was the subject of Example 5-4.) Note that the THL does not include pad or hybrid losses. AT&T calls this a *relative transhybrid loss*. Recall from Example 5-4 that the test termination is removed to establish a *reference THL*. The reference THL is subtracted from the raw THL measurement to obtain the relative THL. Alternatively, the reference THL can be compensated by adjusting the return loss set. Then the relative THL is read directly.

West   For toll applications, AT&T recommends that the ERL of either end be ≥ 33 dB (see Fig. 5-12b). This specification assumes that there is no reflection from the far end. (For testing, this is achieved by using a four-wire CU at the far end, or simply by connecting idle code to the CU under test.)

End-to-End   What minimum echo return loss performance, end to end, can be expected of this trunk?

First assume that each end just meets the AT&T recommendations. The end-to-end ERL will be a function of both the west-end reflection and the east-end reflection. Since the reflections are noise spectra (and one is, furthermore, delayed), they will add on a power basis. Also note that the reflection from the east end receives a round-trip attenuation of 6 dB. The east-end ERL as seen at the west end is

$$34 \text{ dB} + 6 \text{ dB} = 40 \text{ dB}$$

Calculations   We must now combine two ERLs into one end-to-end ERL. First convert each echo return *loss* to a *relative* signal level: 33-dB echo return *loss* (west end) becomes -33 dB relative level, and 40-dB echo return *loss* (east end as seen at west end) becomes -40 dB relative level. We now want a power sum of these two levels. Since they are expressed in decibels, we must first convert each level to a relative power:

$$P_{\text{W}} = \log^{-1}\left(\frac{-33}{10}\right) = 5.01 \times 10^{-4}$$

and

$$P_{\text{E}} = \log^{-1}\left(\frac{-40}{10}\right) = 1 \times 10^{-4}$$

where P_{W} is the power reflected from the west end and P_{E} is the power reflected from the east end as seen at the west end. The total end-to-end echo power is

$$P_{\text{W}} + P_{\text{E}} = 6.01 \times 10^{-4}$$

Expressed in decibels, the relative power is

$$10 \log (6.01 \times 10^{-4}) = -32.2 \text{ dB}$$

Converted back to a *loss*, this is 32.2 dB.

The expected end-to-end ERL performance of this trunk is thus ≥ 32.2 dB. This is not much worse than the near-end performance alone; the ICL of the trunk has diminished the far-end THL contribution.

5-4-2 Balance Networks

Up to this point, we have shown various balance network circuits without being specific about their applications. In this section we present hybrid balance networks by category and discuss their use in the network. The top-level categories have to do with the degree of hybrid balance: precision or compromise.

Precision Balance Networks

A hybrid achieves its greatest degree of balance when the impedance of its balance network equals the impedance connected to its two-wire port. Often the latter impedance is so variable that a close impedance match is not practical. In some cases, however, the impedance at the two-wire port is stable and known. In these cases, high hybrid balance can be achieved by using a *Precision Balance Network* (PBN).

A typical PBN application is shown in Fig. 5-13a. A *hybrid, two-wire repeater* is located at an end office and is used to amplify voice signals on a long cable facility. Since the cable makeup is known and fixed, its impedance is known and relatively stable. (*Cable makeup* refers to the arrangement of wire gauge and load coils used to implement the circuit.) Since we have a good handle on the impedance connected at the two-wire port, we can match that impedance with a precision balance network.

Precision balance networks are available to match various cable makeups and other impedances. Figure 5-13b shows an example circuit for a 24H88 PBN. Note that this network is designed for a one-to-one correspondence with the impedance at the two-wire port. Some hybrid designs, however, call for an impedance-scaled balance network (see Chapter 4). The repeater of Fig. 5-13a could find an application in the telephone network as the repeater in Fig. 5-1b.

Compromise Balance Networks

In most hybrid applications it is not possible to take advantage of the PBNs described above. If the hybrid is associated with a switch, for example, the two-wire impedance may vary from call to call as different circuits are switched to the hybrid. Even if a hybrid is associated with a two-wire circuit on a semipermanent basis, it may be administratively impractical to furnish a different PBN for each circuit.

When PBNs are not feasible, *Compromise Balance Networks* (CBNs) are used. As the name implies, the resulting hybrid balance is a compromise. A handful of CBNs with different applications exist.

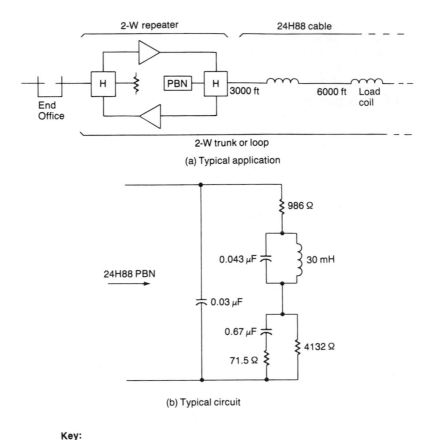

FIGURE 5-13 Precision balance network. (Part (b) adapted courtesy of Siecor Corp.)

Series compromise networks. Before the advent of local digital switching, the only compromise network was what we now call a *series compromise network*. There are two of these networks, the 600- and the 900-Ω compromise networks. Each has a 2.16-μF series capacitor as shown in Fig. 5-14b. In the application in Fig. 5-14a, the hybrid is connected to a different subscriber loop or trunk for each call. In general, the 900-Ω CBN is used at end offices and the 600-Ω CBN at tandem offices. Historically, end offices have an impedance of 900 Ω and toll (now tandem) offices have an impedance of 600 Ω. In the telephone network,

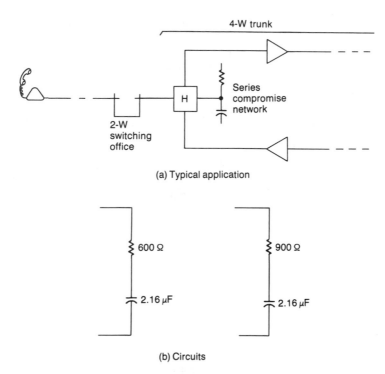

(a) Typical application

(b) Circuits

FIGURE 5-14 Series compromise networks.

series compromise networks would find applications in Fig. 5-1 at hybrids H_1 through H_5.

Parallel networks.   From Fig. 5-14 and the preceding discussion, we see that series compromise networks have been used historically at switches to match trunks and subscriber loops of all cable makeups. This compromise has been satisfactory because of the echo control techniques described earlier in this chapter.

With the advent of local digital switching, there comes a potential singing problem within the office. Refer to Fig. 5-2c and recall that listener echo can lead to singing as the net loss around the loop drops to 0 dB. Now refer to Fig. 5-15a. Here is a digital end office to which the listener echo model can be applied. (Digital offices are inherently four-wire.)

Historically, analog offices have 0-dB cross-office loss. Digital offices should also have 0-dB loss if the network's loss plan is not to be upset. In a digital end office, gain is almost free, so 0-dB cross-office loss is easy to obtain. But in the worst case, this means 0-dB loss around the four-wire path—just the condition for singing. To prevent singing or near-singing, the hybrids must balance well against the subscriber loops. The singing margins obtained using 0-dB loss and series compromise networks are not adequate for *all* loops in an office.

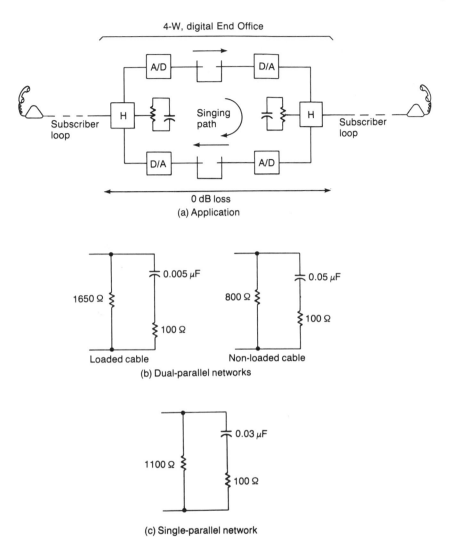

FIGURE 5-15 Parallel networks.

A better balance compromise is obtained by segregating the digital office's loops into a loaded group and a nonloaded group. *Dual-parallel networks* (Fig. 5-15b) are then used, one type for each group. By segregating loops and using dual-parallel networks, 0-dB cross-office loss is permissible. In the telephone network, dual-parallel networks would be used at hybrids H_6 and H_7 of Fig. 5-1.

In situations where circuits cannot be segregated and a single compromise network is needed to match a mixture of loaded loops, nonloaded loops, and trunks, the *single-parallel network* of Fig. 5-15c may be used.

5-4-3 Build-Out Components

In some large, older two-wire analog switching offices, the capacitance of the office wiring can affect the balance of hybrids in trunks connected to the office. To bring the hybrids into balance, a compensating capacitance can be added to the balance network. This is done with a *Network Build-Out Capacitor* (NBOC) (Fig. 5-16a).

As a further complication, the wiring capacitance varies for different path lengths through the office. This varying capacitance can be equalized to some extent by placing *Drop Build-Out Capacitors* (DBOCs) on the two-wire ports of the hybrids (Fig. 5-16a). In summary, the DBOCs make the wiring lengths look

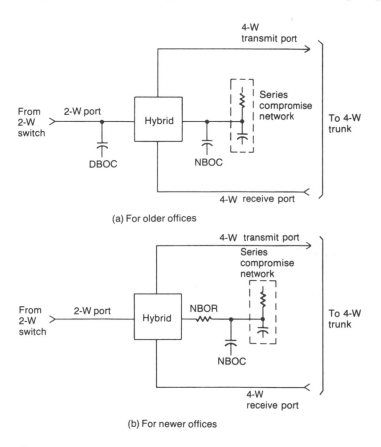

(a) For older offices

(b) For newer offices

Key:
DBOC   Drop Build-Out Capacitor
NBOC   Network Build-Out Capacitor
NBOR   Network Build-Out Resistor

FIGURE 5-16   Build-out circuits. (Adapted from Bellcore *Notes on the BOC Intra-LATA Networks—1986.*[13])

equal, and the NBOCs balance the hybrids for the increased capacitances seen by their two-wire ports.

In newer offices, DBOCs are omitted. It is still necessary to use NBOCs and Network Build-Out Resistors (NBORs) as shown in Fig. 5-16b. The NBORs compensate for office wiring resistance.

An example office that may receive treatment with build-out components is the two-wire tandem of Fig. 5-ld. In other applications, network build-out capacitors can be useful any time a slight adjustment is needed to better balance a hybrid. For example, an NBOC used with a precision balance network could provide an even more precise match than the PBN alone.

5-5 OFFICE BALANCE

5-5-1 Through Balance

Switching office *through balance* is a measure of how well calls route through an office without experiencing reflections. The lower the reflected power (higher return loss), the better the through balance. Through balance is a consideration where Intertandem Trunks (ITTs) are switched together at two-wire tandem offices. Since ITTs are almost all four-wire, four wire-to-two wire conversions are necessary in order to switch these trunks through two-wire offices. (Through balance is not an issue at *four-wire* tandems since calls pass through on a four-wire basis with no reflections.) Recall that an important step for controlling long-distance echo is the minimization of reflections at intermediate offices. Thus the importance of through balance.

Figure 5-17a shows the test setup for measuring through balance. First a path is established through the office between two trunk circuits. (The ITTs associated with these trunk circuits are disconnected.) One trunk circuit is terminated with 600-Ω resistors at its four-wire port. At the other trunk circuit, a four-wire return loss measurement is made. Office balance is specified in terms of return loss, so the measurement can be compared directly with the balance objective.

Good office balance is obtained by adding build-out components to the hybrid balance networks. These build-out capacitors and resistors compensate for office wiring as was discussed above in the section on build-out components. Figure 5-17a shows their use in a two-wire tandem office. In practice, either all or a statistical sample of trunk circuits is measured to verify that an office meets through-balance objectives.

5-5-2 Terminal Balance

Terminal balance is a measure of impedance match between Tandem-Connecting Trunks (TCTs) and the rest of the network. Specifically, terminal balance is a four-wire return loss test made from an ITT appearance at a tandem office. The

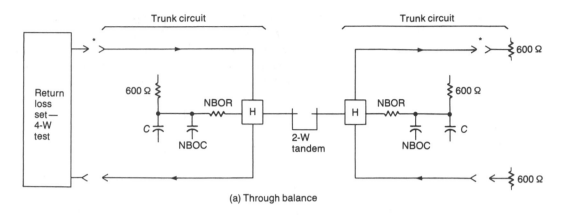

(a) Through balance

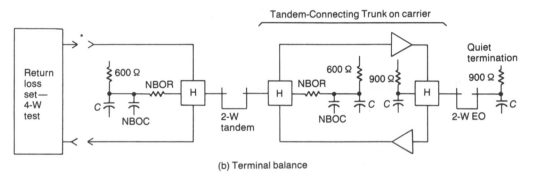

(b) Terminal balance

Key:
C   2.16 μF
NBOR   Network Build-Out Resistor
NBOC   Network Build-Out Capacitor
EO   End Office
2-W   Two-wire
*   Port normally connected to Intertandem
           Trunk

FIGURE 5-17   Office balance. (Adapted from Bellcore *Notes on the BOC Intra-LATA Networks—1986.*[13])

test is made looking through the tandem switch, through the TCT, and through the End Office (EO) toward a quiet termination. Terminal balance is thus a measure of the reflections seen by the far-end ITT on a long-distance call. These reflections will come from the far-end tandem office, the TCT, the end office, and the quiet termination. Good terminal balance is important for echo control since the reflections involved are the predominate echo heard on long-distance calls.

Terminal balance includes the effects of two switching offices and the connecting trunk. Since each of these three elements can be two-wire or four-wire, there are eight basic circuit arrangements for testing terminal balance. One fairly complicated example is shown in Fig. 5-17b. (All eight arrangements are depicted in Bellcore *Notes on the BOC IntraLATA Networks-1986.*[13])

In the figure there is a 600-Ω, two-wire tandem office. If this office also switches intertandem trunks, it will be treated for through balance as shown by the build-out components. Even if the tandem office does not switch intertandem trunks, the hybrid networks may need build-out components in order to meet terminal balance objectives. The end office is a 900-Ω, two-wire switch. At the EO, the series compromise network matches the quiet termination, which is simply another series compromise network. Note that if the tandem office, TCT, and end office are *all* four-wire, terminal balance is not an issue.

It has not been the intent in this section to present detailed instructions for performing and verifying through and terminal balance. However, you should now have a general feeling for the subject and a knowledge of how build-out capacitors and resistors are used. As the network continues to evolve toward all four-wire switching and transmission equipment, office balance will find fewer applications.

5-6 RETURN LOSS OBJECTIVES AND PERFORMANCE

5-6-1 Return Loss Performance for Subscriber Loops

Data on the echo and single-frequency return loss of subscriber loops are available from the Bell System's 1973 Loop Survey.[23] In this survey, the mean ERL was 11.3 dB and the mean SF return loss at 3 kHz was 7.7 dB. The ERL of 95% of main stations was better (higher) than 7.5 dB. Five percent were better than 17.5 dB. The 95th and 5th percentiles for 3-kHz SF return loss fell at 4 dB and 15 dB. In the survey, the mean SF return loss was calculated every 100 Hz within the voice band. The frequency with the lowest mean (worst return loss) was 3.2 kHz—at the edge of the band.

(The term *main stations* as used above is equivalent to subscribers, not loops. A party line may serve two or more main stations on *one loop*. The cable makeup will be slightly different as seen by each subscriber.)

5-6-2 Return Loss Objectives for Trunks

Return loss is not an issue for trunks implemented end-to-end with four-wire facilities, since there are no points of reflection. However, two-wire trunks do have reflections.

The return loss performance of two-wire tandem-connecting trunks is important in the control of echo on long-distance calls. As discussed earlier, terminal balance is a return loss measurement made on TCTs. Terminal balance measurements include the effects of reflections in the tandem and end-office switches plus those in the TCT. Nevertheless, Bellcore's *terminal balance* objectives are presented in Table 5-2 as *return loss* objectives for TCTs. (All the tables of objectives are located at the end of the chapter.)

5-6-3 Return Loss Objectives for Carrier Systems

The return loss objectives for digital carrier systems are listed in Table 5-3. Examples 5-4 and 5-5 demonstrate some of the measurement methods used for verifying these objectives.

5-6-4 Return Loss Objectives for Switching Equipment

The Bellcore and REA return loss objectives for switching equipment are listed in Table 5-4.

5-6-5 Return Loss Objectives for Subscriber Equipment

Return loss objectives for telephone sets, key telephone systems, and PBXs are listed in Table 5-5.

Some of the test conditions listed for PBXs are referred to as "terminal balance" by the Electronics Industries Association (EIA).[4] This *is not* the same terminal balance as that described earlier for tandem-connecting trunks in the BOC intra-LATA networks.

As used by the EIA in referring to PBXs, the expression "through balance" *is* equivalent to through balance described earlier for the BOC's two-wire tandem offices. Note that both on- and off-hook return loss is specified for PBX trunk circuits. This assures good circuit stability (i.e., no singing) for all signaling states on PBX trunks.

TABLE 5-2 RETURN LOSS OBJECTIVES: TANDEM-CONNECTING TRUNKS

Adapted from Bellcore *Notes on the BOC Intra-LATA Networks—1986*[13]				
	Minimum return loss (dB)			
	Echo return loss		Singing return loss	
Condition[a]	Preservice	Immediate action	Preservice	Immediate action
Four-wire facility	22	16	15	11
Two-wire facility				
Analog tandem	18	13	10	6
Digital tandem	18	16	13	11

[a] This measurement is the same as *terminal balance*. Terminal balance is a four-wire return loss measurement made at the tandem end of a Tandem-Connecting Trunk (TCT). The measurement includes the effects of the tandem switch, the TCT, and the End-Office (EO) switch. After switching through the EO, the circuit under test is terminated in either idle code (digital EO) or a quiet termination consisting of 900 Ω in series with 2.16 μF (analog office).

TABLE 5-3 RETURN LOSS OBJECTIVES: CARRIER SYSTEMS

Adapted from *Digital Channel Bank Requirements and Objectives*[12] (PUB 43801)[a]			
Test circuit configuration and CU type[b]	Reference impedance	Far-end termination	Minimum return loss (dB)
Two-wire test on two-wire CUs[c]	600 Ω or 900 Ω in series with 2.16 μF[d]	Four-wire CU (i.e., no reflection)	Echo: 33[e] Singing: 20
Two-wire test on each port of four-wire CU	600 Ω	—	1 kHz SF[b]: 28 300–3000 Hz SF: 23
Four-wire test on two-wire far-end CU[f]	—	600 Ω or 900 Ω in series with 2.16 μF[d]	Echo: 34 Singing: 20

[a] Reprinted with permission of AT&T. Copyright AT&T 1982; all rights reserved.

[b] CU, Channel Unit; SF, Single Frequency.

[c] Connect 25-Ω loop resistance between the channel unit under test and the return loss set. This simulates office wiring.

[d] Use 600 Ω or 900 Ω as appropriate to match the channel's nominal impedance.

[e] The objective stated is for two-wire E&M channel units used at two-wire tandem offices. For other situations, the ERL requirement is relaxed to 28 dB minimum.

[f] This test actually measures THL of far-end, two-wire CU (see Example 5-4).

TABLE 5-4 RETURN LOSS OBJECTIVES: SWITCHING EQUIPMENT

Adapted from Bellcore *Notes on the BOC Intra-LATA Networks—1986*[13]				
	Minimum return loss (dB)			
	Echo return loss		Singing return loss	
Condition	Preservice	Immediate action	Preservice	Immediate action
Four-wire trunk to four-wire trunk through two-wire office[a]	27	21	20	14

From REA 522 (Digital Local Office)[14]				
		Minimum return loss[b] (dB)		
Measuring Port	Terminating port	ERL	SRL-low	SRL-high
Trunk	Trunk	27	20	23
Trunk	Line	24	17	20
Line	Line or trunk	18	12	15

From REA 524 (Analog Common Control Office)[15]			
		Minimum return loss[c] (dB)	
Measuring Port	Terminating port	ERL	SRL
Line or trunk	Line	24	15

[a] This is the same as *through balance* (see the text).

[b] Reference and terminating impedance for two-wire ports is 900 Ω in series with 2.16 μF. Four-wire ports are terminated in 600 Ω.

[c] Measured against a reference impedance of 900 Ω in series with 2 μF, through the switch to a termination of 900 Ω in series with 2 μF. Measured with 20 and 70 mA dc applied to any port under test that normally conducts dc.

TABLE 5-5 RETURN LOSS OBJECTIVES: SUBSCRIBER EQUIPMENT

From EIA-470 (Telephone Instruments)[5]		
Reference impedance	Frequency range (Hz)	Minimum single-frequency return loss[a] (dB)
600 Ω	500–2500 200–3200	7 3.5

From EIA-478 (Key Telephone Systems)[6]
Same as EIA-470, above

From EIA-464 (PBXs)[4]		
Off-hook talking state		
Condition[b,c]	Minimum return loss[d] (dB)	
	ERL	SF in range 200–3200 Hz
Station to station	18	12
Station to four-wire trunk[e]	24	14
Station to two-wire trunk	18	12
Two-wire trunk to four-wire trunk[e]	28	14
Four-wire trunk to four-wire trunk[f]	27	20

On-hook, off-hook nontalking or ground-start states			
Test type	Condition	Frequency or range (Hz)	Minimum single-frequency return loss (dB)
Two-wire[b]	Loop-start and DID CO trunks	200–3200	6
	Ground-start CO trunks	200–3200	2
Two-wire, $Z_{ref} = 600$ Ω in series with 2.16 μF	Two-wire tie trunks	200 500–3200	9[g] 6
Four-wire	Four-wire tie trunks (lossless[h])	200 500–3200	10[g] 6
	Four-wire tie trunks (CTS[i])	200 500–3200	18[g] 14

TABLE 5-5 *(contd.)*

From FCC Part 68.308[i]	
Condition	Minimum return loss
Two- and four-wire tie trunks, on-hook or off-hook nontalking	Same as EIA-464 immediately above for on-hook tie trunks

[a] Measured over the dc loop current range provided by the loop simulator circuit of Chapter 1.

[b] Reference and terminating impedances equal the corresponding port's nominal impedance (i.e., 600 Ω in series with 2.16 μF or any other impedance for which the PBX is designed).

[c] Measured from either end of connection.

[d] On 95% of connections.

[e] Measured with transmission pads switched into four-wire trunk circuit.

[f] Measured with pads switched out.

[g] In the range 200 to 500 Hz, the minimum return losses are continuous functions of frequency as shown in both EIA-464 and FCC part 68.308.

[h] A *lossless* tie trunk has essentially no loss between the interface and the PBX's TLP.

[i] A *CTS* tie trunk has a 3- to 4-dB loss (for a hybrid or *Conventional Terminating Set*) between the interface and the PBX's TLP.

6 LONGITUDINAL BALANCE

This chapter is about balanced telephone circuits and the two modes in which voice-frequency signals are transmitted on them: metallic and longitudinal. When these circuits become unbalanced, coupling between the two modes can occur. Longitudinal balance is the circuit characteristic that defines this coupling. Poor longitudinal balance can result in noise and crosstalk on telephone circuits. In this chapter we show two ways to measure longitudinal balance, and we list industry objectives for this characteristic.

6-1 BALANCED CIRCUITS

Communications circuits often use two conductors that are balanced with respect to ground. One purpose of this balance is to minimize hum and noise pickup from nearby power lines and equipment. Examples of balanced circuits include 300-Ω TV twin-lead, audio circuits in broadcast and recording studios, differential digital line drivers and receivers, and voice- frequency telephone circuits.

Two Circuit Modes

On a balanced circuit, the desirable or *normal-mode* signal appears between the two conductors. The hum and noise or *common-mode* signal appears between each conductor and ground. In telephony, the normal-mode signal is called the *metallic* signal and the common-mode signal is called the *longitudinal* signal.

If the circuit is balanced, the normal-mode signals are isolated from the longitudinal-mode signals. If the circuit becomes unbalanced, signals of one mode may couple into signals of the other mode. The coupling can occur between modes in either direction. Furthermore, the coupling is not necessarily bidirectional.

6-2 LONGITUDINAL-TO-METALLIC COUPLING

The coupling of longitudinal signals into metallic signals can be illustrated by the example of a subscriber loop. Figure 6-1 shows a telephone set connected to a Central Office (CO) via a twisted pair of wires. The pair is one of many inside a cable. The CO presents a balanced load to the line (i.e., $R_1 = R_2$). The cable is constructed so that the resistance of one wire is equal to the resistance of the wire with which it is paired. Also, the capacitance between ground and each wire of a pair is equal. Thus the cable pair is balanced.

Now let an interfering voltage V be coupled to the pair. We have shown V to be capacitively and equally coupled to each side of the pair. The result is a longitudinal voltage V_L that can be measured between ground and either side of the pair.

If the coupling, the cable pair, and the CO termination are not all perfectly balanced, slightly unequal currents will flow in the two conductors. This results in a metallic voltage, V_M, measurable across the pair. In other words, the unbalance has caused some of the longitudinal signal to be converted into a metallic signal. The ratio of V_L to V_M is a measure of the circuit's balance:

$$\text{longitudinal balance (in dB)} = 20 \log \frac{V_L}{V_M}$$

You may be familiar with the term *Common-Mode Rejection Ratio* (CMRR). The CMRR at the input of a balanced circuit is a measure of how well the circuit

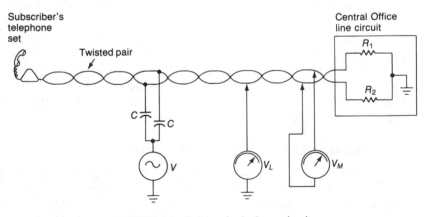

FIGURE 6-1 Balanced telephone circuit.

accepts normal-mode signals and rejects common-mode signals. Common-mode rejection ratio and longitudinal-to-metallic balance are equivalent-circuit characteristics.

Influence, Coupling, and Susceptibility

Influence. Let's discuss the environment surrounding the elements of Fig. 6-1. The source of interfering voltage V is probably a power line near the telephone cable—maybe on the same utility pole. The higher the power line's voltage or current, the higher is V. Engineers concerned with power-line interference refer to V as the *influence*.

The influence is often nonsinusoidal, consisting of the power-line fundamental (usually 60 Hz) plus harmonics. The harmonics are created by nonlinear loads and may extend into the voice band. Minimizing power-line harmonics is a concern of the power company.

Coupling. Next we have some electrical *coupling* from the power line to the telephone line. Figure 6-1 shows capacitive coupling, but there can also be inductive coupling. Coupling is affected by physical separation between power and telephone cables, grounding methods, and telephone cable shielding. Minimizing coupling is the joint concern of power and telephone companies.

Susceptibility. The longitudinal-to-metallic voltage conversion that we have discussed is referred to as *susceptibility*. Recall that if the cable pair is unbalanced, our circuit is susceptible to interference. The manufacturers of telephone cable pay special attention to providing good balance in the conductors. The telephone company is careful to maintain the cable's good balance; for example, installers strive to make low-resistance splices.

Susceptibility is also affected by the longitudinal balance of the equipment that terminates the telephone line. In Fig. 6-1 this equipment is a telephone set at one end and a central office line circuit at the other. By providing good longitudinal balance in terminal equipment such as this, the equipment design engineer can play a part in reducing power-line interference.

Note that all three elements (influence, coupling, and susceptibility) must be present for a power-line interference problem to exist. For example, in the early days of telephony, farmers communicated with each other over grounded one-wire circuits. These unbalanced lines had high *susceptibility*. However, unbalanced lines were acceptable until the rural areas became electrified, at which time the new power lines provided the *influence*.

Measuring Longitudinal-to-Metallic Balance

IEEE method. In its Standard 455-1985,[7] the Institute of Electrical and Electronics Engineers (IEEE) describes a method for measuring longitudinal-to-metallic balance. Figure 6-2 shows the basic circuit for testing one-port equipment such as simple telephones. The figure is deceptively simple. In practice, means

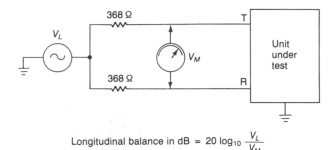

$$\text{Longitudinal balance in dB} = 20 \log_{10} \frac{V_L}{V_M}$$

FIGURE 6-2   Measuring longitudinal balance: IEEE method. (Adapted from ANSI/IEEE Std 455-1985, *IEEE Standard Test Procedure for Measuring Longitudinal Balance of Telephone Equipment Operating in the Voice Band.*[7])

must be provided to calibrate the test set by internally balancing the two 368-Ω resistors. Also, stray test circuit capacitance and inductance must be balanced. Standard 455-1985 specifies the tolerance of the 368-Ω resistors and the degree of internal balance that must be achieved.

The standard also specifies additional test circuitry that must be attached to the second port of two-port equipment. Examples of two-port equipment include cable pairs (where each end is a port) and transformers (where the primary is one port and the secondary is the other).

A means of applying dc bias to the Unit Under Test (UUT) must be provided. A dc bias (such as loop current) is often required to make the UUT function normally. In addition, some longitudinal balance problems may only appear in the presence of dc. When testing telephone equipment, it is best to simulate as much of the normal environment as possible. Applying dc bias is a good example. Note that the bias circuit must not upset the test circuit's internal balance.

The amplitude of V_L should be made variable in order to check for nonlinearities. Since longitudinal balance is usually a function of frequency, the frequency of V_L should be variable over the voice-frequency range.

Test instruments.   When testing a large volume or variety of circuits, you may want to use a commercial test set rather than your own test circuit. An example of a commercial tester is the Wilcom Model T207 Longitudinal Balance Test Set (Fig. 6-3). This instrument has the dc-bias, internal-calibration, dual-port, variable-voltage, and variable-frequency features discussed above as desirable. It also makes measurements that conform to the IEEE standard.

Note that the IEEE does not specify acceptable test results. Standard 455-1985 is merely a *standard method* of making the measurement. When cited in specifications the standard must be accompanied by a test limit; for example: "Longitudinal balance shall be 60 dB or better at 300 to 3000 Hz when measured according to IEEE Standard 455-1985."

Longitudinal-to-Metallic Balance Objectives

When the longitudinal-to-metallic balance of telephone terminal equipment is poor, power-line interference (heard as hum and noise) may become a problem. Balance objectives have been published for equipment at both the subscriber and

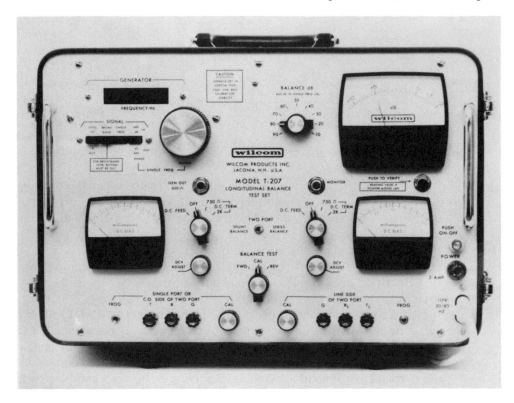

FIGURE 6-3 Longitudinal balance test set. (Courtesy of Wilcom Products, Inc.)

CO end of loops (Table 6-1). Note that the objectives are more stringent (higher objective in dB) at lower frequencies, where power-line harmonics are stronger. (Table 6-1 also includes metallic-to-longitudinal balance, covered in the next section.)

TABLE 6-1 LONGITUDINAL BALANCE OBJECTIVES

From FCC Part 68 (Terminal Equipment)[1]			
Direction	Condition[a]	Frequency (Hz)	Minimum balance (dB)
Metallic to longitudinal	Loop-start, nondata, both on- and off-hook	200–1000 1000–4000	60 40
	Loop-start, data, on-hook	200–1000 1000–4000	60 40
	Loop-start, data, off-hook	200–4000	40
	Ground-start, or reverse-battery; off-hook	200–4000	40

TABLE 6-1 *(cont.)*

From EIA-470 (Telephone Sets)[5] and EIA-478 (Key Telephone Systems)[6]			
Direction	Condition[a]	Frequency (Hz)	Minimum balance (dB)
Longitudinal to metallic	Per IEEE Std. 455, both on- and off-hook[b]	60 600 4000	80 70 54
Metallic to longitudinal	Both on- and off-hook[c]	200–1000 1000–4000	60 40

From EIA-464 (PBX)[4]				
Direction	Condition[a]	Frequency (Hz)	Balance[d] (dB)	
			Minimum	Average
Longitudinal to metallic	Per IEEE Std. 455, off-hook	200 500 1000 3000	58 58 58 53	63 63 63 58
		Frequency[b] (Hz)	Desired minimum balance (dB)	
		60 600 4000	80 70 54	
Metallic to longitudinal	Loop-start CO trunks and loop-start off-premise station line circuits, both on- and off-hook[c]	200–1000 1000–4000	60 40	
	Ground-start CO trunks and reverse-battery CO trunks, off-hook[c]	200–4000	40	

From REA 522 (Digital Central Office)[14]			
Direction	Condition	Frequency (Hz)	Minimum balance (dB)
Longitudinal to metallic	Per IEEE Std. 455, 20- to 70-mA loop current	60–2000 2700 3400	60 55 50

[a] With dc loop current as applicable from loop simulator circuit (see Chapter 1).

[b] Three limit points are shown as derived from the balance versus frequency curve found in EIA-464, EIA-470, or EIA-478.

[c] Measured using FCC method (see Fig. 6-5).

[d] Minimum and average balance taken over 95% of connections.

6-3 METALLIC-TO-LONGITUDINAL COUPLING

So far we have discussed conversion of longitudinal voltages to metallic voltages. This is the susceptibility that gives us power-line noise problems. We now turn our attention to conversion of signals from metallic to longitudinal. Such conversion can cause crosstalk, as illustrated next.

Crosstalk Caused by Poor Longitudinal Balance

Figure 6-4 shows two pairs in the same cable. Each pair connects a CO line circuit with a piece of telephone equipment at a subscriber's location. Let's assume that subscriber A's equipment is slightly unbalanced. We have shown this by placing a capacitor from one side of the pair to ground.

The desirable signal on pair A is a metallic signal. Its origin is V_1. For purposes of illustration, we have shown V_1 coupled to pair A by a transformer. Since the subscriber end of pair A is unbalanced, more current flows in one side of the pair than in the other. This creates a longitudinal voltage on the pair. In other words, there has been a metallic-to-longitudinal conversion of part of signal V_1.

Now, the longitudinal voltage on pair A is converted to a metallic voltage on pair B by interpair crosstalk coupling. (Interpair crosstalk from *longitudinal-to-metallic* modes is about 25 dB worse than the more familiar *metallic-to-metallic* mode crosstalk discussed in Chapter 3.[24]) The undesirable metallic voltage on pair B is measurable as V_2.

In summary, part of a *desired* metallic voltage V_1 on pair A has become an *undesired* metallic voltage V_2 on pair B. If the situation is bad enough, subscriber B will hear subscriber A—a loss of privacy.

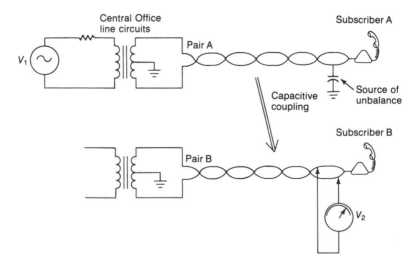

FIGURE 6-4 Crosstalk caused by metallic-to-longitudinal unbalance.

Measuring Metallic-to-Longitudinal Balance

Part 68 of the FCC Rules and Regulations[1] provides a method for measuring the metallic-to-longitudinal balance of telephone terminal equipment. Figure 6-5 is a simplification of the FCC's test circuit. The full test setup includes trim adjustments for achieving high test set internal balance, and a means of supplying dc bias (the dc portion of the loop simulator of Chapter 1 is recommended).

Metallic-to-Longitudinal Balance Objectives

Part 68 of the FCC Rules sets limits on the metallic-to-longitudinal balance of terminal equipment. The basic objective is 60 dB (200 to 1000 Hz) and 40 dB (1000 to 4000 Hz) in *both* the on- and off-hook states. As shown in Table 6-1, this basic objective is relaxed for certain equipment types. For example, the *on-hook* objective is omitted for ground-start trunks since these circuits are inherently unbalanced while on-hook.

In its Part 68 Registration Measurement Guide,[2] the FCC allows the use of the IEEE method of measuring balance—but the performance must be 10 dB better than the limits shown in Table 6-1. Note, however, that the IEEE method measures *longitudinal-to-metallic coupling,* which is not the same as the *metallic-to-longitudinal coupling* specified in Part 68. We advise using caution here. The safest action is to use the measurement method called for in the specification you are using.

Note also that FCC Part 68 sets limits on coupling only in the direction from metallic to longitudinal. Coupling in this direction can cause one subscriber's signals to crosstalk to another subscriber's conversation. Such interference is considered a "harm to the network" and thus controlled by Part 68.

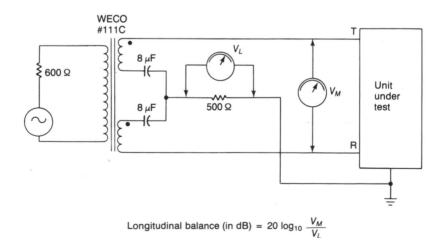

$$\text{Longitudinal balance (in dB)} = 20 \log_{10} \frac{V_M}{V_L}$$

FIGURE 6-5   Measuring longitudinal balance: FCC method.

Coupling in the other direction (i.e., longitudinal to metallic) can cause a subscriber to experience hum and noise, but such coupling does not interfere with *other* subscribers. Thus longitudinal-to-metallic coupling is not covered by FCC Part 68. The FCC has limited its regulation of terminal equipment to matters of network harm, safety, and billing protection. Other performance considerations are allowed to be judged by the marketplace.

6-4 OPERATION IN THE PRESENCE OF HIGH LONGITUDINAL NOISE

Earlier we defined longitudinal balance as the *ratio* between two voltages; we have not yet been concerned with the absolute value of those voltages. No matter how good the longitudinal balance, we cannot raise the longitudinal voltage applied to a unit of telephone equipment beyond its common-mode range.

Example 6-1

Figure 6-6 shows a loop-current detection circuit. Transformer T_1 supplies battery and ground feed to the loop. Note that the ground feed is through resistor R_1. Loop current flowing through R_1 produces a voltage V_1 that is connected to one input of comparator IC_1. When the loop is open (telephone set on-hook), V_1 is at ground. When the loop is closed (telephone off-hook), V_1 goes negative. A threshold voltage V_T is selected such that IC_1 detects on-hook versus off-hook by comparing V_1 with V_T.

We now introduce an interfering longitudinal noise V_L. The resultant longitudinal current in the tip lead creates a noise voltage at V_1. This noise may interfere with the proper detection of on-hook and off-hook states. If V_L is large enough, V_1 may exceed IC_1's common-mode input range; IC_1 may even be damaged.

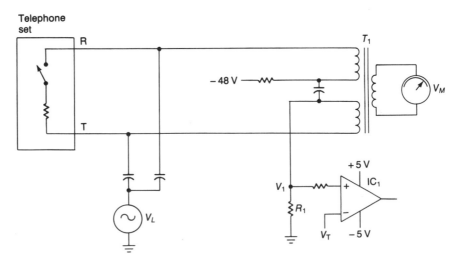

FIGURE 6-6   Operation in the presence of high longitudinal noise.

Note that the circuit in the example may have good longitudinal balance (i.e., no noise at V_M), but may nevertheless cease to function due to high longitudinal noise. In addition to specifying longitudinal balance, standards often require that equipment function correctly in the presence of a few volts applied longitudinally.

Circuits that need evaluation for potential problems in this area include on-hook/off-hook detectors, dial pulse detectors, ring-trip detectors, and surge protection networks.

6-5 FAR-END LONGITUDINAL SUSCEPTIBILITY

Longitudinal voltages can cause noise at the far end of a circuit even though the equipment has good longitudinal balance. We have coined the term *far-end longitudinal susceptibility* to label this phenomenon. Figure 6-7 shows how it occurs,

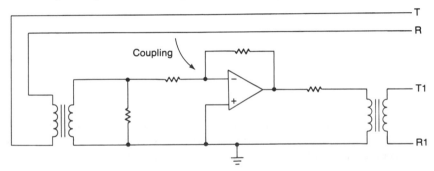

FIGURE 6-7   Far-end longitudinal susceptibility.

using a one-way repeater as an example. Both of the repeater's transformers have good balance. If we measure longitudinal balance (either metallic to longitudinal or vice versa) at leads T and R or T1 and R1, we get good results. However, if we apply a longitudinal voltage to T and R, we get a metallic noise voltage at T1 and R1. In our example, this is due to capacitive coupling on the Printed Circuit Board (PCB) from lead R to a high-impedance node of the amplifier. (Moving around some PCB signal traces clears this problem.)

Note that you must specifically test for far-end longitudinal susceptibility by applying a longitudinal voltage to one end of the circuit and looking for metallic noise at the other end. Far-end longitudinal susceptibility is not detected by either the IEEE or FCC methods of measuring longitudinal balance.

7 ═══════
═══ DISTORTION AND MISCELLANEOUS IMPAIRMENTS

In this chapter we begin with a discussion of several types of distortion. For distortion to occur, a *desired signal* must be present. If this signal is then altered in some undesirable way, we say that it is *distorted*. Distortion can be classified as linear or nonlinear.

Nonlinear distortion is dependent on the signal amplitude. (Think of a hi-fi amplifier that begins to produce distortion as its operating level increases toward the clipping point.) Examples of nonlinear distortion include harmonic, intermodulation, and quantizing distortion—all to be defined later. Nonlinear distortion is caused by transformers, active devices, analog-to-digital converters, and so on. These devices are found in equipment such as repeaters, telephone sets, and voice multiplexers. A characteristic of nonlinear distortion is that signal components are created at frequencies that did not exist in the input signal.

Linear distortion is signal independent. An example of this type of distortion is *amplitude distortion* (referred to as the *frequency response* characteristic in this book—see Chapter 2). In a linear circuit, the frequency response is the same at both low and high operating levels. Other examples of linear distortion include phase and amplitude variations that occur in filters. A characteristic of linear distortion is that the phase and amplitude of the desired signal change without the creation of new frequency components.

The chapter ends with a collection of miscellaneous impairments, including jitter, hits, and dropout. The impairments discussed in this chapter generally have more effect on voice-band data than on speech. If these impairments are controlled for data purposes, they will probably go completely unnoticed by voice users.

7-1 DISTORTION

7-1-1 Harmonic Distortion

Harmonic distortion is nonlinear distortion that creates unwanted frequency components at harmonics of the input signal. For example, if an input at 1 kHz undergoes harmonic distortion, the output will include *distortion products* at 2 kHz, 3 kHz, 4 kHz, and so on. The 2-kHz component is called *second-harmonic distortion* and the 3-kHz component is called *third-harmonic distortion*. All the distortion products taken together are called *total harmonic distortion*. The input frequency does not have to be 1 kHz, but it should be low enough to have harmonics within the passband of the equipment under test.

In telephony, harmonic distortion is expressed as the ratio in dB between the fundamental and the distortion product(s) of interest (second, third, or total).

Example 7-1

If at the output from the equipment under test, the fundamental (e.g., 500 Hz) is -10 dBm and the component at 1500 Hz is -53 dBm, the third-harmonic distortion is 43 dB.

Second- and third-harmonic distortion can be measured by tuning a Frequency-Selective Voltmeter (FSVM) to the frequencies of interest. Total harmonic distortion is measured using an instrument that notches out the fundamental and measures everything (within reason) that is left.

7-1-2 Intermodulation Distortion

Intermodulation distortion is nonlinear distortion that occurs when there are two or more frequency components in the input to the equipment under test. Intermodulation (IM) distortion creates distortion products at frequencies equal to the sums and differences of the input frequencies and to the sums and differences of the harmonics of the input frequencies. To illustrate this, assume that the input frequencies are f_A and f_B. The *second-order IM distortion* products are at $f_A + f_B$ and $f_B - f_A$. The *third-order IM distortion* products are at $2f_A + f_B$, $2f_A - f_B$, $2f_B + f_A$, and $2f_B - f_A$. On telephone circuits, second-order IM distortion is represented by the power average of the $f_A + f_B$ and $f_B - f_A$ components. For third-order IM distortion, it is sufficient to measure only the $2f_B - f_A$ component.

The measurement of IM distortion is preferred over that of harmonic distortion when assessing nonlinearities in Voice-Frequency (VF) telephone circuits. For simple circuits with only one source of distortion, the results of the two measurements would be similar. However, typical telephone channels comprise many sources of distortion both linear and nonlinear. On such channels, IM distortion provides a truer assessment of non-linear distortion effects.

Four-Tone Method

In its standard 743-1984,[8] the Institute of Electrical and Electronics Engineers (IEEE) specifies a *four-tone method* for measuring IM distortion on telephone circuits. Instead of using two tones (frequencies), the four-tone method uses two *tone pairs* where the tones of each pair are closely spaced. The four tones are equal in level and are transmitted at a composite level of -13 dBm0. The test tone levels and frequencies have been selected so that nonlinear distortion will affect the test signal in a manner similar to the way it affects data signals. We will discuss the frequency selection in more detail later.

The four test tones are at frequencies $f_{A1} = 857$ Hz, $f_{A2} = 863$ Hz (the first pair), $f_{B1} = 1372$ Hz, and $f_{B2} = 1388$ Hz (the second pair). The tones are shown in Fig. 7-1 along with the multitude of second- and third-order distortion products that may result.

Recall that the second-order products we wish to measure are at $f_A + f_B$ and $f_B - f_A$. Since our input consists of two tone *pairs* (not two single tones), there are actually *eight* second-order products. The $f_A + f_B$ products fall in a band centered at 2240 Hz. Their individual frequencies are $f_{A1} + f_{B1}$, $f_{A1} + f_{B2}$, $f_{A2} + f_{B1}$, and $f_{A2} + f_{B2}$. The $f_B - f_A$ products fall in a band centered at 520 Hz. Their individual frequencies are $f_{B1} - f_{A1}$, $f_{B1} - f_{A2}$, $f_{B2} - f_{A1}$, and $f_{B2} - f_{A2}$.

Also recall that the third-order product we wish to measure falls at $2f_B - f_A$. The use of tone pairs here results in *six* third-order products in a band centered at 1900 Hz. Their individual frequencies are $2f_{B1} - f_{A1}$, $2f_{B1} - f_{A2}$, $2f_{B2} - f_{A1}$, $2f_{B2} - f_{A2}$, $f_{B1} + f_{B2} - f_{A1}$, and $f_{B1} + f_{B2} - f_{A2}$.

Selection of test frequencies. The four input frequencies used in the four-tone IM distortion method may at first seem odd choices. Actually, they were selected with care. The spacing within the pairs was made unequal (6 Hz versus 16 Hz). If this had not been done, the two second-order products at $f_{A1} + f_{B2}$ and at $f_{A2} + f_{B1}$ would fall on the same frequency. These two products would add on a voltage basis, producing erroneous results.

The test frequencies selected allow the IM bands of interest to fall within the flattest part of the channel's frequency response curve (assumed here to be 500 to 2300 Hz). Review Fig. 7-1 to see that this is so.

Finally, if the input tones experience jitter in traversing the channels, the jitter components will usually fall within 300 Hz of each input tone. Therefore, the input frequencies are selected so that the IM distortion products of interest fall at least 300 Hz from any input tone. This can also be verified by examining Fig. 7-1.

These criteria for selecting the input frequencies have been made so that other impairments will have minimal effect on the IM distortion measurement.

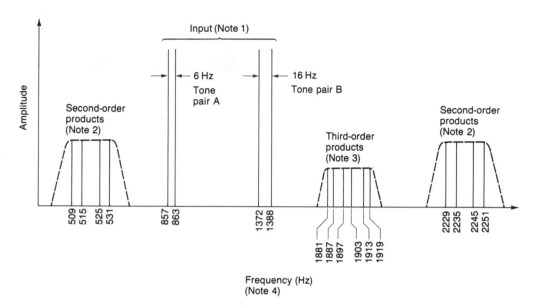

FIGURE 7-1   Four-tone intermodulation distortion measurement: spectrum of test and distortion frequencies.

NOTES: 1. The four equal-level input tones have a total power level of −13 dBm0.

2. Four second-order distortion products are measured via a 50-Hz bandwidth filter centered at 520 Hz. Another four second-order products are measured via a 50-Hz bandwidth filter centered at 2240 Hz. The total second-order distortion level is the power average of the outputs of these two filters.

3. Six third-order distortion products are measured via a 50-Hz bandwidth filter centered at 1900 Hz.

4. The horizontal axis is not to scale.

Instruments for Measuring Intermodulation Distortion

The circuitry for measuring intermodulation distortion is usually included in multipurpose test instruments such as the one shown in Fig. 7-2. This instrument will measure basic transmission parameters plus several impairments that we introduce in this chapter.

Figure 7-3 is a functional block diagram of typical IM distortion measurement circuitry. The transmit block and the receive block would usually be in two separate instruments, one at each end of the channel under test. The two blocks could also be in the same instrument when measuring a looped-around channel or an isolated piece of equipment. The output and input circuitry provides impedance matching, balanced sources and terminations, talk/dial access, and hold functions as described earlier for other instruments.

In the transmit block, the four equal-level test tones are normally mixed and applied to the circuit under test at data level (−13 dBm0 for the composite signal

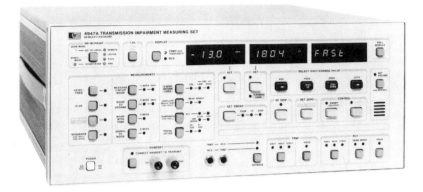

FIGURE 7-2 Transmission impairment measuring set. (Courtesy of Hewlett-Packard Co.)

or −19 dBm0 for each tone). In the receive block, the two second-order bands (centered at 520 and 2240 Hz) and the third-order band (centered at 1900 Hz) are separated by three 50-Hz wide bandpass filters. The rms voltage outputs from the 520-Hz and 2240-Hz filters (V_{S1} and V_{S2}) represent the second-order IM distortion. Voltages V_{S1} and V_{S2} are power-averaged to become V_S. The circuitry in the power average block computes

$$V_S = \left(\frac{V_{S1}^2 + V_{S2}^2}{2}\right)^{1/2}$$

(Notice that we are discussing signal *voltage* and not *power*. This follows the practice of engineers who think in terms of voltage while designing the internal stages of equipment. The internal voltage level is converted to a power level at the input and output terminals. This conversion is necessary since the VF interface between pieces of equipment is specified in terms of power.)

Continuing with the IM distortion receive block, a switch selects between the second- and third-order distortion products, V_S and V_T. The selected IM product (V_{IM}) is compared with the rms input voltage (V) to determine the Signal-to-Distortion ratio (S/D), where S/D (in dB) = 20 log (V/V_{IM}).

Although we use rms voltages in the calculations above, the instrument's detector may be rms, quasi-rms, or average responding.

Noise check. The four-tone method reduces the effect of steady noise on the measurement by using the most narrow of filters (50 Hz) sufficient to capture the distortion products. However, a noisy channel may still cause the IM distortion to appear worse than it is. The transmit block of Fig. 7-3 contains a *noise check* switch, which is used to test for this condition. When the noise check switch is operated, one tone pair is removed from the transmitted signal and the level of the other tone pair is increased 3 dB (to maintain the same transmitter

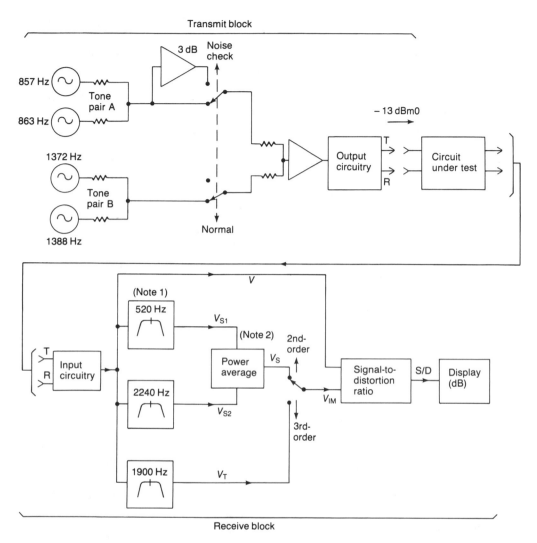

FIGURE 7-3 Four-tone intermodulation distortion measurement: block diagram of instruments.

NOTES: 1. The three filters have 50-Hz bandwidths.
2. See text for equations relating the voltages and the S/D ratio.

signal power). With one of the tone pairs absent, there are no IM distortion products falling within the passbands of the three 50-Hz filters. Thus the receive block detects only noise and its display shows the Signal-to-*Noise* ratio (S/N). If the S/N ratio is greater than the S/D ratio by at least 12 dB, the noise can be ignored. If the difference is less than 1 dB, there may be excessive noise which should be investigated. For situations between these two limits, the noise influence on the IM distortion measurement can be negated by compensating the reading as discussed next.

In *normal* mode, the instrument actually measures the Signal-to-*Noise plus Distortion* ratio [S/(N + D)]. In *noise check* mode, the instrument measures the *S/N* ratio—we want the *S/D* ratio. The distortion power equals the noise-plus-distortion power minus the noise power. Using this relationship and the principles from Chapter 2, we could calculate the S/D ratio from the S/(N + D) and S/N ratios. Newer instruments will do this calculation for you. You can also use the noise compensation graph found in the manuals for older instruments.

Range and accuracy.   Instruments that comply with the IEEE's standard for four-tone IM distortion measurement must have an S/D accuracy of ±1 dB and a measurement range from 10 to 70 dB.

Patent protection.   The four-tone IM distortion measurement method is protected by U.S. Patent 3,862,380, assigned in 1975 to Hekimian Laboratories, Inc.[27] Other manufacturers who use this method in their equipment do so under license from Hekimian.

Intermodulation Distortion Objectives and Performance

Table 7-2 at the end of this chapter shows IM distortion objectives for various parts of the network. Individual network parts are allowed less distortion than the network as a whole. Carrier systems, for example, are allowed little IM distortion. Since a number of carrier systems may connect in tandem to provide an end-to-end link, the impairments contributed by each system must be small. The relatively high total network objective is indicative of many distortion sources adding along the end-to-end link.

Second- and third-order IM distortion products are specified separately. The third-order products are expected to be lower in amplitude, yielding higher signal-to-distortion ratios.

For IM distortion performance, we look to the Bell System's 1982/83 End Office Connection Study (EOCS).[22] This survey used the four-tone method to measure IM distortion in the predivestiture network. Measurements were made on built-up connections from local CO to local CO. The results show that IM distortion is not closely related to distance.

The mean IM S/D ratios for both second- and third-order products, for all calls in the study, was 53 dB. Only 0.3% of calls had performance worse than the 27-dB second-order and 32-dB third-order limits set for the total network.

7-1-3 Quantizing Distortion

A common means for transmitting telephone conversations is to convert the voice to a digital signal. Once digitized, the voice signal is less susceptible to impairments in passing through switching offices and transmission links. However, digitization creates its own problems.

Figure 7-4 shows a simplified view of the digitizing process. At the top is shown an analog signal containing a continuous range of amplitudes. This signal is sampled at intervals as shown by the tick marks. Between sampling points, the amplitude is *held* at the sample amplitude (Fig. 7-4b) so that it can be digitized. In the North American Pulse Code Modulation (PCM) system, 8 bits are available for encoding each sample. This yields 127 discrete amplitudes on the positive side of zero and 127 on the negative side. Zero amplitude is encoded as one of two equivalent codes: $+0$ or -0. There are then 255 discrete amplitudes for encoding the voice signal.

In the encoding process, each sample is forced to one of these 255 amplitudes. That is, the continuous range of input amplitudes is *quantized* into 255 discrete amplitudes. Figure 7-4c shows the amplitudes assigned to the input signal when it is quantized. The difference in amplitudes between the signal in Fig. 7-4a and the signal in Fig. 7-4c is called the *quantizing error*. It is plotted in Fig. 7-4d. The quantizing error is a function of the input signal and the system's parameters (such as the number of bits per sample). If there is no input, there is no quantizing error. Thus the error signal is a distortion product. The ratio of the input signal to the C-message-weighted quantizing error is called the *quantizing distortion*.

Quantizing Distortion Measurement

Quantizing distortion can be measured with fairly simple instruments: a test oscillator and a noise measuring set equipped with notch and C-message filters.

Example 7-2

Figure 7-5a shows the measurement of quantizing distortion, end to end, on a D-channel bank. For this example we picked 600-Ω, four-wire channel units with -16 TLP transmit and $+7$ TLP receive ports. AT&T's specifications for digital channel banks (Table 7-3) call for measurements to be made over a range of input levels and at each of three frequencies. Our example shows only the measurement made at 0 dBm0 and 1004 Hz.

Recalling that the transmit port is at -16 TLP, we insert a 1004-Hz sine wave at a level of -16 dBm. At the receiving end, the channel unit is connected to a Noise Measuring Set (NMS). The 1010-Hz notch filter is switched in to reject the 1004-Hz tone. The residual noise and distortion is then C-message weighted and (in simpler sets) displayed as an absolute power in dBrnC.

Suppose that the reading is 63 dBrnC. We must convert this to a signal-to-distortion ratio since that is how quantizing distortion is defined. Since the receive port is at $+7$ TLP, the absolute *distortion* reading of 63 dBrnC is equivalent to 56 dBrnC0. The *signal* level of 0 dBm0 is equivalent to 90 dBrnC0. The S/D ratio is

$$90 \text{ dBrnC0} - 56 \text{ dBrnC0} = 34 \text{ dB}$$

This is 1 dB better than the end-to-end limit of 33 dB shown in the specification.

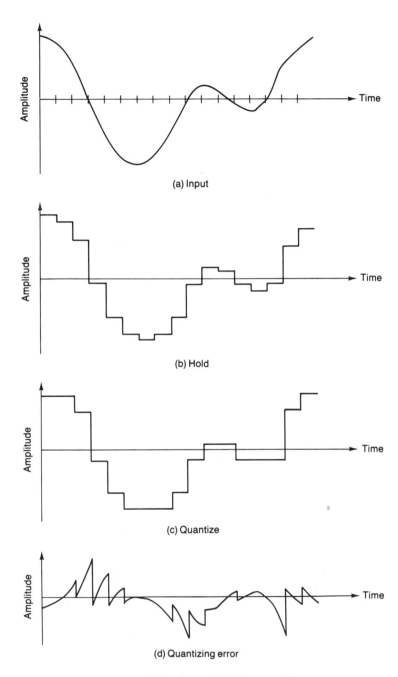

FIGURE 7-4   Quantizing distortion: waveforms.

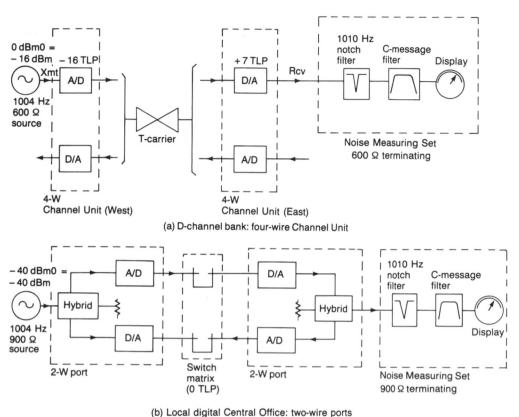

(a) D-channel bank: four-wire Channel Unit

(b) Local digital Central Office: two-wire ports

FIGURE 7-5   Quantizing distortion: measurement examples.

Noise measuring sets more sophisticated than the one in the preceding example may handle some or all of the ratio calculations for you. For example, the NMS may compensate for a nonzero TLP and it may display the S/D ratio directly.

Example 7-3

Figure 7-5b shows the measurement of the cross-office quantizing distortion of a local digital central office. We will illustrate a measurement made at −40 dBm0. Since the ports are at 0 TLP, −40 dBm0 equals −40 dBm. We apply a 1004-Hz sine wave at that level to the input port. A NMS is connected to the output port and its 1010-Hz notch and C-message filters are switched in. Assume that the NMS reads 26 dBrnC. What is the S/D ratio?

The method chosen for calculating the S/D ratio is a variation of the method used in Example 7-2. Since all ports are at 0 TLP, there is no need to convert to dBrnC0. The signal and distortion levels must, however, be in the same units. To convert the signal unit (dBm) to match the distortion unit (dBrnC), add 90 dB. The S/D ratio is then

$$-40 \text{ dBm} + 90 \text{ dB} - 26 \text{ dBrnC} = 24 \text{ dB}$$

If we were testing to the switching equipment objectives of Table 7-3, this measurement would not meet the limit of 27 dB.

Quantizing Distortion Objectives and Performance

Quantizing distortion is specified for individual pieces of equipment such as digital carrier channel banks and digital switching offices. The manufacturers of this equipment control its performance by design. Upon installation, the performance is verified by the user. Unless the equipment becomes defective, it will probably continue to meet objectives for a long time.

Quantizing distortion objectives are usually stated for an end-to-end connection or a cross-office connection. Such a connection includes one coder-decoder (codec) at each end. This measurement is made between analog points-using simple instruments as illustrated in Fig. 7-5. More sophisticated instruments are available that allow quantizing distortion to be measured on a single codec. This single-end-only measurement is made between an analog point and a digital point. The Hewlett-Packard 3776B PCM Terminal Test Set is an example of an instrument with this capability.

Table 7-3 lists objectives for both end-to-end and single-end-only quantizing distortion. Notice that the end-to-end objective is the same for all the equipment types listed. The quantizing distortion standards for North American PCM are quite similar regardless of the equipment type. The manufacturers of single-chip codecs also design to this standard. These Integrated Circuits (ICs) may be used in equipment ranging from small key systems to large toll offices. In essence, there is a single objective (and performance) for quantizing distortion on links comprising two codecs.

Such links in *tandem,* however, create a different situation. Some built-up connections may contain several A/D and D/A converters (coders and decoders). Impairments such as noise and quantizing distortion are additive at each conversion. When the standards for PCM were developed, this additive effect was considered. Small impairments were allocated to each link, so that a connection comprising many links would still be acceptable. As digital networks evolve, there will be fewer connections with multiple A/D and D/A conversions. Users of these networks will enjoy very quiet connections with no more noise and distortion than was previously allocated to a single link.

Quantizing distortion is usually not measured at the network level. Instead, C-notched noise (which includes the effects of quantizing) is measured on built-up connections.

7-1-4 Aliasing Distortion

This section covers the phenomenon of *aliasing* and the peculiar forms of distortion and noise that it causes. Aliasing occurs in sampled data systems such as those used when voice is digitized for transmission or switching. Aliasing causes a desired input signal of one frequency to be converted to another, undesired frequency. This is called *aliasing distortion* or *foldover distortion*. Aliasing can also convert an undesired, but harmless, noise into noise at a more disturbing frequency. This is called *aliasing noise*. The discussion starts with a somewhat idealized but adequate review of sampled data systems.

Sampled Data Systems: General

In a *sampled data system* the signal is sampled at intervals and information about each sample is transmitted through the system. (This is in contrast to continuous signal transmission.) For analog signals, the samples may be sent in analog form or they may be digitally encoded. Samples from multiple channels may be interleaved (or multiplexed) and transmitted on one circuit. The two different pulse code modulation systems used in voice telephony and on compact discs are examples of sampled data systems that digitally encode and multiplex analog signals.

The time domain.   A typical analog signal is shown as waveform A in Fig. 7-6a. (This figure is a graphical aid that can be used to explain sampled data systems without resorting to heavy use of math.) Waveform A is sampled regularly with a period T. In the time domain, sampling is accomplished by multiplying the input signal with a *sampling waveform*—waveform B in Fig. 7-6a. (In the *time domain,* signal *waveforms* are plotted as functions of time.) Notice that the sampling waveform is an infinite series of narrow pulses, spaced apart at intervals of time T. (In the ideal, pulses that are infinitesimally narrow in time are known as *impulses.*)

Graphically, multiply waveform A by waveform B and see that the result is waveform C. Waveform C is a series of pulses at intervals T. The amplitude of each pulse is proportional to the amplitude of the analog input at the instant the input is sampled.

In a digital system, the samples would be encoded, transmitted (or switched) to the far end, then decoded. The block diagram in Fig. 7-6b shows this process. (Note, however, that digital encoding/decoding is not a requirement in sampled data systems.) Our goal is to recover waveform A at the output of this sampled data system. However, the output developed so far is waveform C, which does not look like waveform A. It may not be clear how to proceed if we are limited to thinking in the time domain.

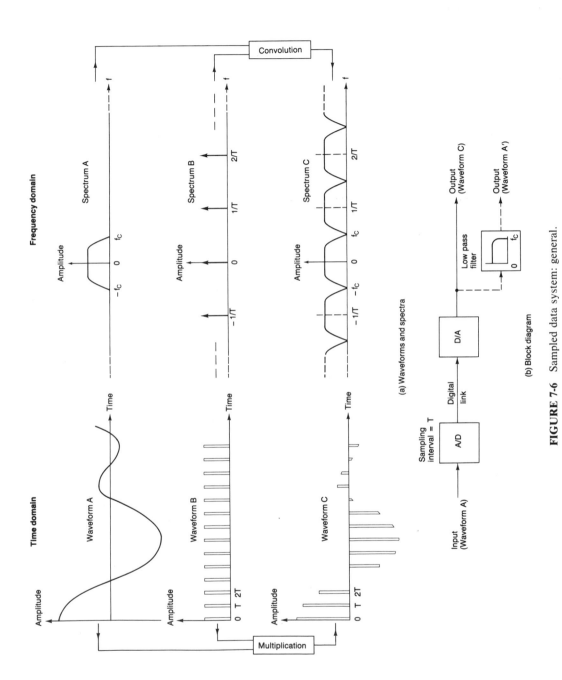

(a) Waveforms and spectra

(b) Block diagram

FIGURE 7-6 Sampled data system: general.

186

The frequency domain. Enter the *frequency domain* where signal *spectra* are plotted as functions of frequency. All signals can be expressed in both the time and frequency domains. That is, all signals can be plotted as functions of time or as functions of frequency. Once a signal is expressed as a function of either time or frequency, it can be converted to a function of the other by using a mathematical technique called a *transform*. (Using logarithms to multiply by performing addition is an example of using a transform.) The *Fourier transform* converts a signal from the time to the frequency domain. The *inverse Fourier transform* converts in the opposite direction.

Spectrum A in Fig. 7-6a is the frequency-domain representation of waveform A. We have defined this signal to contain frequency components between 0 Hz and frequency f_C. (As a mathematical convenience it is customary also to show negative frequencies. That is, for each frequency component f, there is a component of equal amplitude at frequency −f. The introductory chapters in Brigham, *The Fast Fourier Transform*[18] contain good background material on the concepts introduced here.)

Spectrum B (Fig. 7-6a) is the Fourier transform of waveform B. You probably know that a sharp-edged waveform contains an infinite series of progressively higher frequency components. Waveform B is a pulse that repeats with period T. Thus in the frequency domain it has a fundamental frequency of 1/T. Since waveform B consists of infinitesimally narrow (i.e., sharp) pulses, its spectrum contains all harmonics of 1/T. Waveform B has a dc component which transforms to the impulse at 0 Hz in spectrum B. Finally, spectrum B contains matching positive and negative frequency components. Examining Fig. 7-6, we see that the Fourier transform of a series of impulses at intervals T (waveform B) is another series of impulses at intervals 1/T (spectrum B).

We continue now with spectrum C, which is the Fourier transform of waveform C. However, we will derive spectrum C not from waveform C, but from spectra A and B. Recall that in the time domain, waveform C was obtained by multiplying waveform A by waveform B. Multiplication in the time domain is equivalent to *convolution* in the frequency domain. We can convolve the input (spectrum A) with the sampling function (spectrum B) to obtain the sampled data output (spectrum C).

Fortunately for our discussion, the convolution of a spectrum with an impulse is a simplified case of general convolution. To convolve spectra A and B, simply replicate spectrum A at each impulse in spectrum B. The result (spectrum C) is a series of energy bands of width $2f_C$ centered at frequencies 0, 1/T, 2/T, and so on.

In the frequency domain, the output (spectrum C) still does not look like the input (spectrum A). However, we can see that spectrum A is contained within spectrum C (examine spectrum C for the band from $-f_C$ to $+f_C$). How to recover the original input from output C is now clear. Send the output through a low-pass filter with cutoff just above f_C. This is shown by the dotted connection in Fig.

7-6b. This low-pass output filter is called the *postsampling filter* or the *receive filter*.

Sampled Data System: PCM

The PCM system used in North American telephony is a specific example of a sampled data system. Continuing in the frequency domain, Fig. 7-7 shows some of the characteristics of this system. Figure 7-7a is an example analog test input. The dotted line shows the VF range, while the discrete spectral line indicates the presence of a 1-kHz tone. The input is sampled at intervals of 125 μs (a sampling rate of 8 kHz) (Fig. 7-7b). The output (Fig. 7-7c) is the convolution of Fig. 7-7a and b. There are now discrete spectral lines at frequencies located 1 kHz above and 1 kHz below all the harmonics of 8 kHz (including 0 Hz and the negative harmonics).

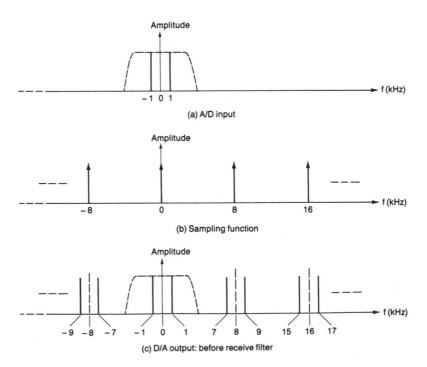

FIGURE 7-7 Sampled data system spectra: North American PCM standard. (Part (d) adapted from *Digital Channel Bank Requirement and Objectives* (PUB 43801). Reprinted with permission of AT&T. Copyright AT&T 1982; all rights reserved.)

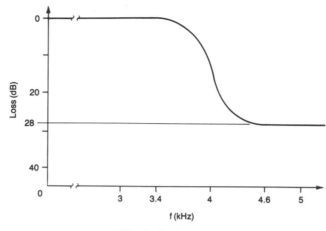

(d) Receive filter loss characteristic

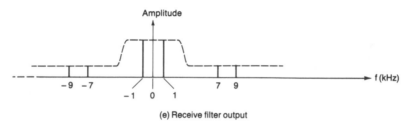

(e) Receive filter output

FIGURE 7-7   (*cont.*)

This multitude of frequencies (the *sampling products*) passes through the receive filter (Fig. 7-7d). The receive filter is quite flat up to 3400 Hz, then rolls off to 14 dB loss by 4000 Hz, and to 28 dB or greater loss above 4600 Hz. This filter characteristic will attenuate all the unwanted sampling products by at least 28 dB. The final output is shown in Fig. 7-7e. The output is a restoration of the input waveform, distorted slightly by the residual components at 7 kHz, 9 kHz, and so on. We will show later how the amplitudes of these components are measured in order to assess the frequency response of the receive filter.

Aliasing

Consider now the spectrum in Fig. 7-8a. This input signal is a 5-kHz tone. The sampling function (Fig. 7-8b) is the same as before. The sampling process convolves the input and sampling functions to yield the spectrum of Fig. 7-8c. As before, this spectrum has components at 5 kHz above and below all harmonics of

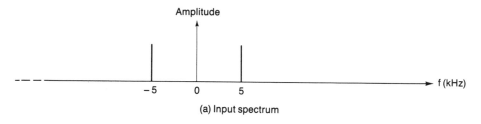

(a) Input spectrum

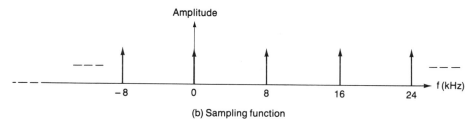

(b) Sampling function

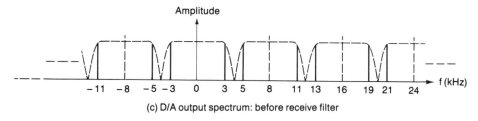

(c) D/A output spectrum: before receive filter

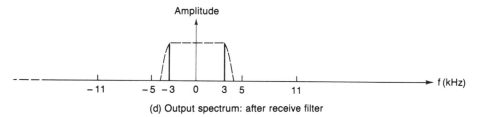

(d) Output spectrum: after receive filter

FIGURE 7-8   Aliasing distortion.

the 8-kHz sampling frequency. For example, the component 5 kHz below the first harmonic falls at

$$1 \times 8 \text{ kHz} - 5 \text{ kHz} = 3 \text{ kHz}$$

The signal next passes through the receive filter whose response is shown dotted in Fig. 7-8c. The 3-kHz component passes with no attenuation, and the components at 5 kHz and above are attenuated at least 28 dB (Fig. 7-8d). Note that we have applied a 5-kHz tone to the input of the sampled data system, and have received a 3-kHz tone at its output (compare Fig. 7-8a and d). This phenomenon is called *aliasing*.

Aliasing occurs when the input frequency is greater than the *Nyquist* frequency. The Nyquist frequency is equal to one half of the sampling frequency. The Nyquist frequency is 4 kHz in telephony PCM. In the example, aliasing occurred because the sampling frequency was not at least twice the input frequency. Viewed another way, the input frequency was too high for the standard sampling rate of 8 kHz. To prevent aliasing, the input must be *bandlimited* to 4 kHz. This is accomplished by using another low-pass filter, called the *presampling, antialiasing,* or *transmit filter* (Fig. 7-9). The frequency responses of the transmit and receive filters are similar; however, the transmit filter has slightly more stop-band attenuation ($\geq$32 dB above 4.6 kHz). In telephony, transmit filters also have a low-frequency roll-off, not shown. (The purpose of the low-frequency roll-off is to attenuate power-line interference.)

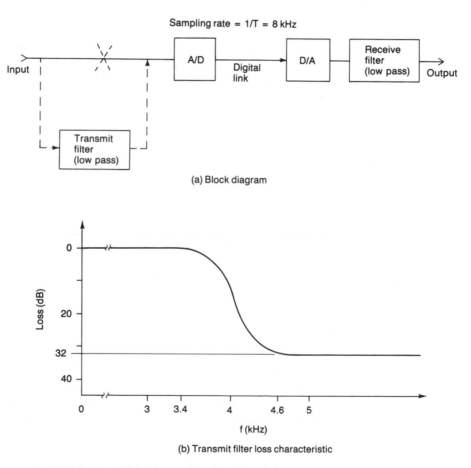

(a) Block diagram

(b) Transmit filter loss characteristic

FIGURE 7-9   PCM system. (Part (b) adapted from *Digital Channel Bank Requirements and Objectives* (PUB 43801). Reprinted with permission of AT&T. Copyright AT&T 1982; all rights reserved.)

Single-Frequency Distortion Objectives

For PCM systems, the sampling product and aliasing distortion residuals are lumped together and covered by a *single-frequency distortion* objective (Table 7-4). This table lists wide ranges of frequencies over which tests are to be made. Based on our new knowledge of sampled data systems, we can select a smaller set of input test frequencies and also predict at which output frequencies the residuals will fall. The techniques are presented in the next section.

Single-Frequency Distortion Measurement

Figure 7-10a shows the test setup for measuring single-frequency distortion on PCM systems.

Example 7-4

Figure 7-10b shows a measurement made using an *in-band* input frequency (3 kHz). The 3-kHz tone appears at point A and is plotted with the transmit filter response superimposed. The tone passes through the transmit filter to appear unattenuated at point B.

Next, the signal is sampled and all the sampling products appear at point C. Recall that these products are tones located 3 kHz above and below all harmonics of 8 kHz (and 0 Hz). The first few tones are plotted in the figure.

The out-of-band sampling products are next attenuated by the receive filter. Upon using the FSVM to examine the spectrum for point D, we will expect to find the original input tone at 3 kHz and at a level of 0 dBm0. We will also expect to find sampling components at 5 kHz, 11 kHz, 13 kHz, and so on, each at a level of -28 dBm0 or less.

In the example above, an *in-band* input tests the stop-band response of the *receive* filter.

Example 7-5

Figure 7-11 shows a measurement made using an *out-of-band* input tone (6 kHz at 0 dBm0). (Refer to Fig. 7-10a for the circuit locations of points A to D.) The 6-kHz tone is attenuated at least 32 dB by the transmit filter.

After sampling, the sampling products appear at point C. The products fall at 6 kHz above and below 0 Hz and at 6 kHz above and below all harmonics of 8 kHz. Notice in particular the aliasing product at $8\text{ kHz} - 6\text{ kHz} = 2\text{ kHz}$. The receive filter will attenuate the sampling products above 4 kHz, but it will not attenuate the 2-kHz aliasing product, which falls in-band. The spectrum for point D shows what we would expect to measure on the FSVM. The important test here is that the 2-kHz tone is at a level of -32 dBm0 or less (as determined by the transmit filter).

In this example, an *out-of-band* input tests the stop-band response of the *transmit* filter.

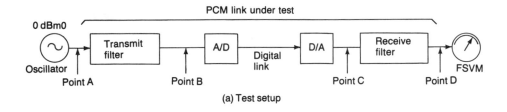

(a) Test setup

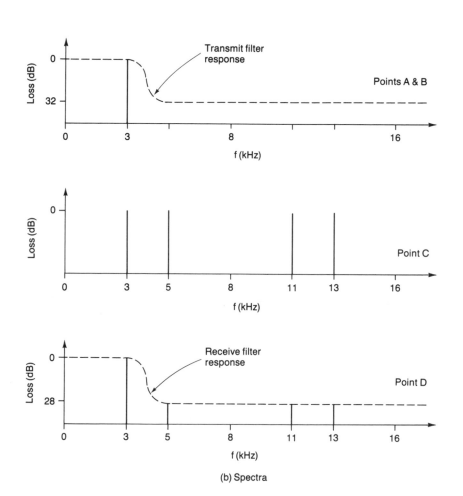

(b) Spectra

FIGURE 7-10   Measuring single-frequency distortion: inband input.

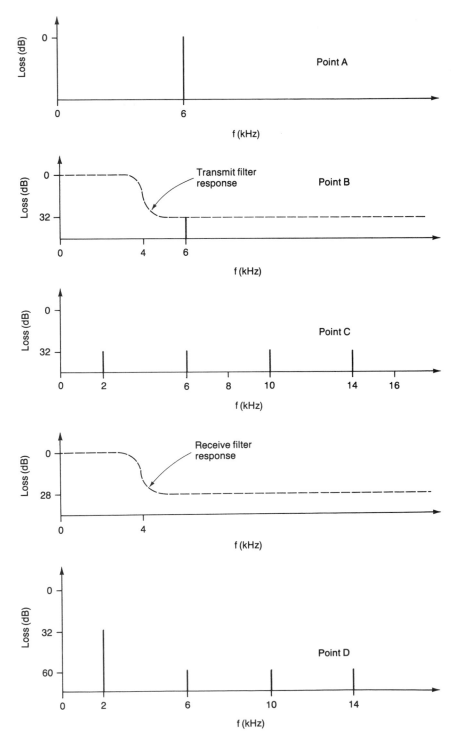

FIGURE 7-11 Measuring single-frequency distortion: out-of-band input.

In summary of the filter characteristics:

1. In-band signals (up to 3.4 kHz) are passed without attenuation by both transmit and receive filters. The out-of-band sampling products from in-band signals are attenuated at least 28 dB by the receive filter.
2. Out-of-band signals (above 4.6 kHz) are attenuated at least 32 dB by the transmit filter.
3. Input signals at frequencies between 3.4 and 4.6 kHz are attenuated by both filters. The attenuation sum in this band is at least 28 dB.

Now review Table 7-4 to see how this summary is reflected in the single-frequency distortion objectives.

Choice of test frequencies. The transmit and receive frequency response curves shown in this section are not curves for any particular filters. They are instead templates that show the minimum stop-band attenuation allowed. Actual filters have more loss but may also have ripple, peaks, and valleys in the stop-band. To fully test a filter's response, you should pick input test frequencies whose sampling and aliasing products fall at peaks of minimum filter loss. This will verify that the filter has the required 28 or 32 dB loss at even the worst-case frequencies.

Aliasing Noise

Aliasing noise was described briefly in Chapter 3, but a fuller discussion has had to wait until aliasing itself was explained. *Aliasing noise* originates at a benign (i.e., inaudible) frequency, but is converted to noise at a harmful frequency by the aliasing process.

Example 7-6

Figure 7-12 shows portions of two desktop communication devices. These devices contain both video display and digitized voice circuitry. Assume that a poor circuit board design allows a 15.734-kHz horizontal video signal to couple into the voice path at a point after the transmit filter.

Such a high-frequency tone would often go unnoticed since it would be greatly attenuated by most analog voice paths. However, this is a sampled data system and the offending tone has bypassed the antialiasing (transmit) filter. The tone produces an aliasing product at

$$2 \times 8 \text{ kHz} - 15.734 \text{ kHz} = 266 \text{ Hz}$$

This low-frequency tone passes unattenuated through the receive filter and may be annoyingly audible at the receiving end.

In recent designs, the transmit filter and A/D converter may be combined into one IC (shown dotted in Fig. 7-12). Nevertheless, high-frequency noise can

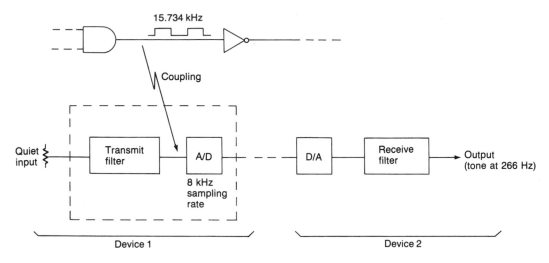

FIGURE 7-12   Aliasing noise.

still bypass the transmit filter and couple via the power supply or other pins directly into the A/D converter.

Aliasing noise is not limited to single-frequency tones. Bands of high-frequency noise can be converted via aliasing into bands of lower-frequency noise (heard as hiss).

7-2 MISCELLANEOUS IMPAIRMENTS

7-2-1 Jitter

In this section we cover phase jitter and amplitude jitter—rapid deviations from nominal in the phase and amplitude of a test tone. These impairments can affect data transmission but have little effect on voice. Of the two, phase jitter is the more important and commonly made measurement.

Phase Jitter

Deviation in the zero crossings of a tone, relative to their nominal location, is called *phase jitter*. This term aptly describes the image (Fig. 7-13a) of zero crossing jittering in time due to the ac content of the deviation. Phase jitter can affect the performance of voice-band modems that use phase modulation. Phase jitter has little effect on voice transmission due to the ear's relative insensitivity to phase information.

There are two influences that can each cause the zero crossings of a signal to jitter: incidental (unwanted) phase modulation, and noise components. Phase

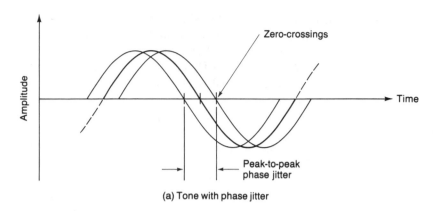

(a) Tone with phase jitter

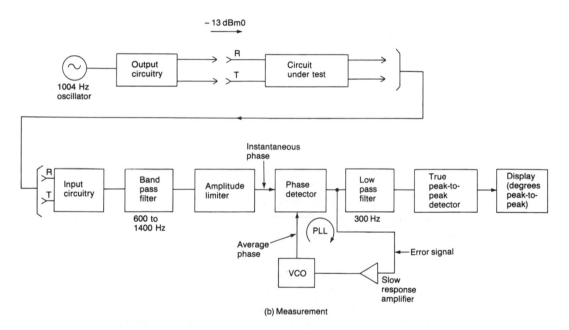

(b) Measurement

FIGURE 7-13   Phase jitter. (Adapted from Operating Manual for Hewlett-Packard Model 4945A Transmission Impairment Measuring Set.)

Modulation (PM) can occur on Frequency-Division Multiplex (FDM) systems if the local oscillator output is contaminated by ringing supply or commercial power frequencies. The PM would contain components at 20 Hz, 60 Hz, and their second through fifth harmonics (20 to 300 Hz).

Phase jitter measurement.   Phase jitter is measured using a 1004-Hz holding tone sent at data level (-13 dBm0). At the receiving end, the jitter in the tone's zero crossing is measured. Figure 7-13b shows the test setup and functional block diagram for a phase jitter meter. The input and output blocks contain circuitry previously described. The bandpass filter limits the instrument's response to the band 400 Hz on either side of the 1004-Hz carrier. This reduces the effect of noise on the measurement. Next the signal is amplitude-limited and presented to one input of a phase detector. This signal is the *instantaneous phase*. The phase detector, a Voltage-Controlled Oscillator (VCO), and a slow-response amplifier form a Phase-Locked Loop (PLL). The VCO output is the *average phase* of the holding tone. The phase detector compares the instantaneous and average phases, then outputs their difference as an *error signal*. The error signal controls the VCO, but with a slowed response. This response is such that the VCO cannot track jitter above 20 Hz. The error signal (which is the same as the jitter) traverses a 300-Hz low-pass filter before being peak detected and displayed. Thus jitter components between 20 and 300 Hz are measured. The measurement range is 0 to 25° peak-to-peak (p-p) as specified by IEEE standard 743-1984.[8] (At 1004 Hz, 25° is equivalent to 69 μs.)

The jitter range 20 to 300 Hz is referred to as *Bell standard* and is a requirement of IEEE standard 743-1984. As an option, the standard allows jitter measurement in the band 4 to 20 Hz (referred to as *low frequency*) and in the band 4 to 300 Hz (referred to as *Bell standard plus low frequency*). The Transmission Impairment Measuring Set (TIMS) shown in Fig. 7-2 is typical of sets that include phase jitter among their capabilities.

Recall that phase jitter measurements include the effects of noise. For example, a channel with 3.3 kHz of white Gaussian noise at a level corresponding to a 30-dB S/N ratio contributes about 3° p-p jitter to the total measurement.[11] The PCM quantizing noise of a 1004-Hz tone also contributes about 3° p-p jitter to the measurement.[11] If a phase jitter measurement is high, this may be caused by noise and not PM. To check this, C-notched noise should also be measured. If C-notched noise is okay, there is probably a PM problem.

Phase jitter objectives and performance. Phase jitter objectives set by Bellcore and AT&T are shown in Table 7-5.

The 1982/83 EOCS provides phase jitter performance data showing this impairment to be correlated with distance. When phase jitter was measured in the band 2 to 300 Hz, the first and 99th percentiles for short calls were 2° and 18° p-p. For medium-distance and long calls, the first and 99th percentiles worsened to 3° and 23° p-p. The mean for all calls was 6° p-p.

When the EOCS performance is compared with Bellcore's objective of 15° p-p, roughly 95% of all calls would have met the objective.

Amplitude Jitter

Deviation in the peak amplitude of a tone relative to its nominal amplitude is called *amplitude jitter*. This jitter contains ac components and would appear on an oscilloscope as shown in Fig. 7-14a. Amplitude jitter can cause errors in high speed modems.

Total amplitude jitter is the sum of two components: incidental (unwanted) amplitude modulation, and noise and interference. Incidental Amplitude Modulation (AM) is the less likely of these two to occur.

Amplitude jitter measurement. Amplitude jitter is measured using a 1004-Hz holding tone sent at data level (−13 dBm0). At the receiving end, the peak

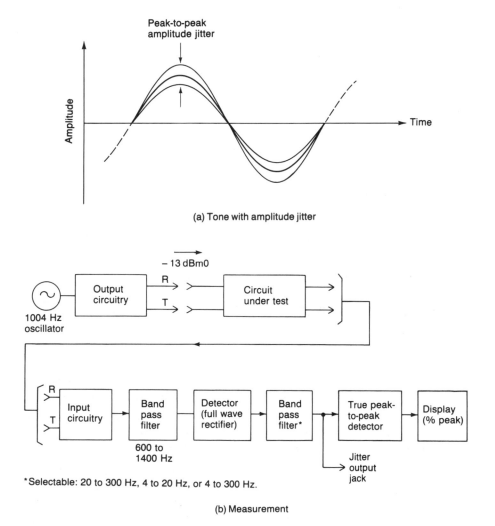

(a) Tone with amplitude jitter

*Selectable: 20 to 300 Hz, 4 to 20 Hz, or 4 to 300 Hz.

(b) Measurement

FIGURE 7-14   Amplitude jitter. (Adapted from Operating Manual for Hewlett-Packard Model 4945A Transmission Impairment Measuring Set.)

jitter in the tone's amplitude is measured. Figure 7-14b shows the test setup and functional diagram for an instrument that measures amplitude jitter. The first bandpass filter limits the instrument's response to a band extending 400 Hz on either side of the 1004-Hz carrier. This reduces the effect of noise on the measurement.

Next, the envelope of the holding tone is detected by the rectifier and second bandpass filter. This detected envelope is the demodulated amplitude jitter signal and may be accessible for analysis on an external jack. The second bandpass filter has the same three selectable response bands as those of the phase jitter meter: Bell standard (20 to 300 Hz), low frequency (4 to 20 Hz), and combined (4 to 300 Hz). Finally, the jitter signal is peak-to-peak detected and displayed. The IEEE standard 743-1984 specifies a display range of 0 to at least 25% peak. (A 25% peak amplitude jitter represents an amplitude swing of about 4 dB in the level of the holding tone.) The ability to measure amplitude jitter is found in TIMS such as the HP 4945A and the Hekimian Laboratories model 3701 with the appropriate option installed.

Since amplitude jitter measurements include the effects of noise, C-notched noise should be measured along with amplitude jitter. If the noise measures high, noise is probably the cause of the jitter—not amplitude modulation.

Amplitude jitter performance. The 1982/83 EOCS measured amplitude jitter in the band 2 to 300 Hz. The first and 99th percentiles for all call distances were roughly 2% and 10% peak amplitude jitter. The mean for all calls was about 3% peak.

7-2-2 Hits and Dropout

Hits and dropout are transients. When they occur in voice-frequency channels, they can affect modem signals. They normally have little effect on voice.

Gain Hits

A *gain hit* is a sudden change in the level of a signal. Following the hit, the level may return to its original value or it may stay at its new value indefinitely. Figure 7-15a shows a number of gain hits affecting a tone. Gain hits can be caused by faulty equipment or by normal actions such as the manual or automatic reconfiguration of circuits along a built-up connection.

Gain hit measurement. Gain hits are usually measured simultaneously with three other transients: impulse noise (covered in Chapter 3), phase hits, and dropout (both covered in following sections).

Figure 7-16 shows the test setup and instrument functional diagram for measuring multiple transients. Referring to the figure, the input and output circuitry match impedances and provide balanced terminations (with dc paths if required to hold the circuit under test). At the measuring end, a bandpass filter limits the set's

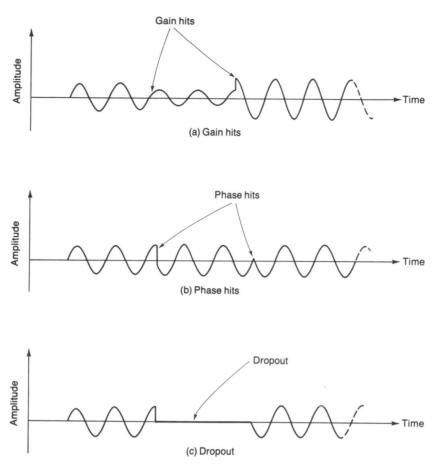

FIGURE 7-15   Hits and dropout. (Adapted from Operating Manual for Hewlett-Packard 4945A Transmission Impairment Measuring Set.)

hit response to a band centered around 1000 Hz. Next, gain hits above an adjustable threshold are detected in a circuit that contains an Automatic Gain Control (AGC). The AGC tracks the holding tone level, thus providing an average reference level with which the instantaneous level can be compared. Rapid level changes that qualify as hits are sent to the logic circuitry. The logic sorts out simultaneous transients by blocking gain hits occurring within 1 second after a dropout. If the logic decides that a gain hit is to be counted, a signal is passed to the gain hit counter. All the counters run simultaneously for an adjustable time; then the counts are displayed. The TIMS pictured in Fig. 7-2 contains the multiple-transient measuring function just described.

The IEEE's standard 743-1984 specifies the performance of gain hit test sets. These sets measure gain changes that are both above and below the nominal level

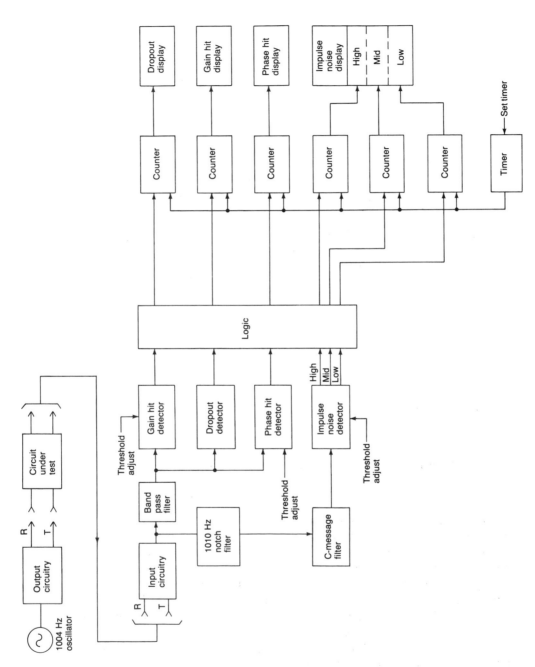

FIGURE 7-16 Measuring transients: hits, dropout, impulse noise. (Adapted from Operating Manual for Hewlett-Packard Model 4945A Transmission Impairment Measuring Set.)

of the 1004-Hz holding tone. The gain-change threshold can be set to 2, 3, or 6 dB; optional additional settings should be at 4, 8, and 10 dB. Gain changes that last more than 4 ms are counted as hits. The count rate is specified as 8 counts per second maximum.

The IEEE also specifies the characteristics of the detector's AGC. This specification is in terms of the *rate of change* of the received level. A rapid change is counted as a hit; a slowly changing level is not. Quantitatively, a linear level change (plus or minus) of 4 dB in 200 ms should be counted as a 2-dB gain hit. A slower change of 4 dB in 600 ms should not be counted.

Gain hit objectives and performance.   The network objective for gain hits is listed in Table 7-6. Data from the 1982/83 EOCS indicate that about 99.3% of all calls met the objective of 8 counts or less in 15 minutes above a threshold of 3 dB.

Phase Hits

A *phase hit* is a sudden change in the phase of a signal. Following the hit, the phase may return to its original value or it may stay at its new value indefinitely. Figure 7-15b shows a number of phase hits affecting a tone. Just as with gain hits, phase hits can be caused by defective equipment or by automatic and manual patching of circuits for maintenance purposes.

Phase hit measurement.   Recall from the preceding section that phase hits are usually measured simultaneously with other transients. Phase hit measurement (Fig. 7-16) is similar to that for gain hits. For phase hits, a separate detector, counter, and display are used.

The phase hit detector contains a Phase-Locked Loop (PLL). The PLL tracks the holding tone phase, thus providing an average reference phase with which the instantaneous phase can be compared. As with gain hits, the logic blocks the counting of phase hits for 1 second following the detection of a dropout.

The IEEE's standard 743-1984 covers phase hit measuring functions. Phase hit test sets measure changes both positive and negative relative to the nominal phase of the 1004-Hz holding tone. Hit thresholds are provided at 5° increments from 5 to 45°. Phase changes lasting longer than 4 ms are counted as hits. The counter's maximum rate is limited to 8 counts per second.

The IEEE also specifies the PLL characteristics of the phase hit detector. The specification is in terms of the *rate of change* of the instantaneous phase relative to the nominal phase. A linear phase change (plus or minus) of 100° in 20 ms is considered a rapid change and should be counted as a 20° phase hit. A slower phase change of 100° in 50 ms should not be counted.

Phase hit objectives and performance.   Table 7-6 lists the network objectives for phase hits. Performance data are available from the 1982/83 EOCS. In that survey, about 98.5% of all calls met the objective of 8 hits or less in 15 minutes above a phase hit threshold of 20°.

Dropout

A *dropout* is a reduction of 12 dB or more in the level of a 1004-Hz holding tone (Fig. 7-15c). This impairment is a special case of gain hit (i.e., a dropout is a *negative* gain hit exceeding 12 dB). A dropout may last for a short time (just over 4 ms) or it may last indefinitely. Conditions such as equipment malfunction and high, weather-related attenuation on microwave links can cause dropout.

Dropout measurement. A multitransient instrument (such as previously described) measures dropout simultaneously with other transients. Referring to Fig. 7-16, the dropout function has its own detector, counter, and display. Unlike the gain hit detector, there is no AGC in the dropout detector. Instead, the received level is measured and stored at the start of the measurement interval. This stored level becomes the reference with which the instantaneous level is compared. If a dropout is detected, the logic blocks the counting of the other transients for 1 second. The dropout counter is limited to a count rate of one per second.

Dropout performance. The network's dropout performance was measured in the 1982/83 EOCS. Roughly 85% of calls had no dropouts during the 15-minute measurement interval. Another 14% had between 1 and 10 dropouts, and the remaining 1% of calls had greater than 10 dropouts in 15 minutes.

Summary of Transients

Table 7-1 summarizes the major parameters used when measuring hits, dropout, and impulse noise. The transients are listed in order of frequency of occurrence; impulse noise occurs most frequently, followed by hits and dropout. In

TABLE 7-1 SUMMARY OF TRANSIENTS

Transient	Measurement threshold	Transient duration for valid count (ms)	Maximum counter rate (counts/s)	Counting hierarchy
Impulse noise[a]	Low: 30–110 dBrn Mid: 4 dB above low High: 4 dB above mid	<4	8	Impulse noise, gain, and phase hits can all count simultaneously if not blocked by a dropout. Noise impulses above a higher threshold will also be counted at all lower thresholds.
Gain hit	2–6 dB[b] (plus or minus)	>4	8	
Phase hit	5–45°[c]	>4	8	
Dropout	−12 dB[b]	>4	1	A dropout blocks counting of all other transients for 1 second.

[a] Metallic; see Chapter 3 for more information on impulse noise.

[b] Relative to level of received 1004-Hz holding tone.

[c] Relative to phase of received 1004-Hz holding tone.

contrast, dropouts are the greatest cause of modem errors, followed by hits, which are followed by noise impulses.

7-2-3 Frequency Shift

Frequency shift (also called *absolute frequency offset*) is simply a change in frequency incurred by a signal as it passes through a channel. This impairment causes all of the signal's frequency components to be shifted by an equal amount.

Frequency shift can be created in carrier systems that use Single-Sideband (SSB) modulation. In SSB systems the carrier is not transmitted but is reinserted at the receiving end as part of the demodulation process. If the reinserted carrier is off in frequency, the demodulated signal will suffer a frequency shift equal to the difference between the modulating and demodulating carrier frequencies. When bad enough, this phenomenon creates the talking-duck effect well known to amateur radio operators.

Incidents of SSB-caused frequency shift should be on the decrease as old, inaccurate carrier supplies are replaced. Frequency shift is also reduced by synchronizing carrier systems to a *frequency synchronization network*. Such a network distributes a reference frequency to central offices. Frequency synchronization networks exist for both analog and digital transmission systems.

Frequency shift can cause errors in modems that divide a channel into narrow bands as part of the modulation process. Imagine a modem sending tones that are received at the far end by using narrow bandpass filters. If the modem's signal is frequency shifted, the tones may fall too far away from their assigned frequencies for proper detection.

Frequency shift measurement.   Frequency shift is measured by sending a known frequency from one end of a circuit and measuring the frequency received at the other end using a frequency counter. The *absolute* frequency accuracies of the sending oscillator and receiving counter must be better than the desired measurement accuracy. The measurement should also be averaged for several seconds to minimize the effects of noise on the counter. A receiving-end bandpass filter centered around 1 kHz and located just before the counter is also helpful. The IEEE's standard 743-1984 requires that frequency shift be measured using an oscillator stable to ±0.005 Hz and a counter accurate to ±0.1 Hz. At 1 kHz and in the presence of noise, the counter is allowed 30 seconds to reach the required accuracy.

Note that no reference frequency is established at the receiving end. Also, there is little point in measuring frequency shift on a looped-around channel since the shifts in each direction may cancel each other.

Frequency shift objectives and performance.   Table 7-7 lists frequency shift objectives for the telephone network. In the 1982/83 EOCS, no conclusions were drawn about the network's performance with regard to frequency shift. The EOCS did find this impairment present (with shifts of $+2$ Hz and -1 Hz) on one route that used SSB modulation. Bell's earlier 1969/70 Connection Survey found 2 calls out of 600 with shifts greater than 3 Hz. From these limited data, it would appear that the network meets its frequency shift objectives.

7-2-4 Incidental Modulation

Incidental modulation is unwanted amplitude, frequency, or phase modulation of a signal by some external influence. In telephony, incidental modulation is usually not measured directly. Instead, we measure jitter since this is easier. Amplitude jitter includes amplitude modulation and noise; phase jitter includes phase modulation, frequency modulation, and noise. Thus the way to check for incidental modulation in practice is as follows:

1. Measure jitter.
2. Measure noise.
3. If the jitter is high and the noise is low, incidental modulation is probably present.
4. If both jitter and noise are high, incidental modulation may or may not be present.

7-3 REFERENCE TABLES

TABLE 7-2 INTERMODULATION DISTORTION OBJECTIVES[a]

Carrier System Objectives from AT&T PUB 43801, *Digital Channel Bank Requirements and Objectives* (1982)[12]		
Condition	Order of distortion products	Minimum signal-to-distortion ratio (dB)
End to end	Second Third	50 54

Subscriber Equipment Objectives from EIA-464 (PBXs)[4]		
Connection type	Order of distortion products	Minimum signal-to-distortion ratio[b] (dB)
Station to station	Second Third	40 43
Station to trunk and trunk to trunk	Second Third	45 53

TABLE 7-2 (*cont.*)

Condition	Order of distortion products	Minimum signal-to-distortion ratio (dB)
Total Network Objectives from Bellcore *Notes on the BOC Intra-LATA Networks—1986*[13]		
Intra-LATA connection (subscriber to subscriber)	Second Third	27 32

[a] Using four-tone method as described in the text. Measured at −13 dBm0.

[b] Must be met by 95% of connections.

TABLE 7-3 QUANTIZING DISTORTION OBJECTIVES

Input level (dBm0)	Minimum signal-to-distortion ratio[b,c]	
Carrier System Objectives Adapted from *Digital Channel Bank Requirements and Objectives*[12] (PUB 43801)[a]		
	End to end (dB)	Transmit or receive only[d] (dB)
0 to −30	33	35
−40	27	29
−45	22	25

Input level[e] (dBm0)	Minimum signal-to-distortion ratio[c,f] (dB)
Switching System Objectives from REA 522 (Digital Central Office)[14]	
0 to −30	33
−30 to −40	27
−40 to −45	22

[a] Reprinted with permission of AT&T. Copyright AT&T 1982; all rights reserved.

[b] Measured at three input frequencies: 1004 Hz (within the range 1004–1020 Hz is acceptable), 304 Hz, and 2996 Hz.

[c] C-message weighted.

[d] Not applicable at 304 Hz or 2996 Hz.

[e] At 1004 or 1020 Hz.

[f] The S/D ratio is allowed to be 2 dB worse (lower) if the least significant bit is lost by the digital switch.

TABLE 7-4 SINGLE-FREQUENCY
DISTORTION OBJECTIVES[a]

Adapted from *Digital Channel Bank Requirements and Objectives*[12] (PUB 43801)[b]	
Input frequency range	Maximum output[c]
0–4600 Hz	−28 dBm0
4600–12,000 Hz	−32 dBm0

[a] Includes aliasing distortion and sampling products.

[b] Reprinted with permission of AT&T. Copyright AT&T 1982; all rights reserved.

[c] Measured end to end. At any input frequency in the range shown and at an input level of 0 dBm0, the output at *any other frequency* shall not exceed the level specified.

TABLE 7-5 PHASE JITTER OBJECTIVES

From Bellcore *Notes on the BOC Intra-LATA Networks—1986*[13]		
Condition	Jitter frequency band	Maximum jitter (peak to peak)
End-to-end intra-LATA call	20–300 Hz 4–300 Hz	10° 15°
Individual carrier terminal	Not specified	1.3°

From AT&T PUB 41004, *Data Communications Using Voiceband Private Line Channels* (1973)[10]		
Condition	Jitter frequency band	Maximum jitter (peak to peak)
Leased channels[a]	20–300 Hz	10°

[a] As provided by the predivestiture Bell System.

TABLE 7-6   GAIN AND PHASE HIT OBJECTIVES

From *Transmission Systems for Communications*, 5th ed., Bell Laboratories (1982)[16]			
Count period	Impairment	Threshold	Maximum counts
15 minutes	Gain hits	3 dB	8
	Phase hits	20°	8

TABLE 7-7   FREQUENCY SHIFT OBJECTIVES

From AT&T PUB 41004, *Data Communications Using Voiceband Private Line Channels* (1973)[10]	
Condition	Maximum frequency shift
Leased channels[a]	±5 Hz

[a] As provided by the predivestiture Bell System.

8

STANDARDS
ORGANIZATIONS
and
REFERENCES

In this chapter we list the more important of the U.S. organizations that publish standards, specifications, and rules related to telephony. Grouped by organization are annotated references cited in the text. In addition to standards and rules, these references include books and technical articles.

Documents such as the FCC rules have the force of law. Customer-provided equipment cannot be sold unless it complies with the rules. Other documents, such as the EIA series, contain specifications that are strongly recommended but not absolutely required. A third category of standard simply defines a method for making a measurement, with no performance objectives stated. Examples include some of the IEEE standards. Finally, there are the field surveys of performance, usually reported in technical journal articles and conference papers. These studies of actual performance can be used to refine the objective-setting standards.

We have provided sources and prices for the standards. The prices are approximate, since some organizations add shipping and handling costs and some offer member discounts.

8-1 FEDERAL COMMUNICATIONS COMMISSION

The Federal Communications Commission (FCC) sets rules that include the regulation of terminal equipment connecting to the telephone network. The FCC rules are available from the Government Printing Office (GPO) as noted below.

[1]"Connection of Terminal Equipment to the Telephone Network," Federal Communications Commission Part 68, *Title 47 Code of Federal Regulations,* Oct. 1, 1987 (Washington, DC: Office of the Federal Register National Archives and Records Administration). Part 68 of the FCC rules is a legal document. It states the federal rules regulating the technical characteristics of customer-provided telephone equipment. Its main purpose is to prevent harm to the telephone network and to telecommunications workers. In general, FCC Part 68 does not specify equipment performance (i.e., how good it sounds). Topics covered include voltage surge, leakage current, hazardous voltages, excessive signal levels, longitudinal balance, on- and off-hook impedance, Ringer Equivalent Numbers (REN), billing protection, and hearing aid compatibility. A bound volume of Title 47 Code of Federal Regulations (CFR) containing FCC Parts 40 through 69 is available for $10.00 from the Superintendent of Documents, U.S. Government Printing Office, Washington, DC 20402. The GPO accepts charge card sales at 202-783-3238. Part 68 contains 147 pages.

[2]"Part 68 Registration Measurement Guide," Office of Science and Technology (OST) Bulletin No. 51, Dec. 1979 (Washington, DC: Federal Communications Commission). Includes illustrated test setups and sample data for many of the FCC Part 68 requirements.

[3]"Instructions for Completing FCC Form 730," June 1988 (Washington, DC: Federal Communications Commission). Form 730 is the application for registration of telephone equipment in accordance with Part 68 of the FCC rules. The instructions include a variety of practical tips pertaining to FCC registration plus a list of test laboratories. The form and its instructions are revised from time to time and are available from: Federal Communications Commission, 2025 M Street, NW, Washington, DC 20554. Phone 202-632-7000.

8-2 ELECTRONICS INDUSTRIES ASSOCIATION

The Electronics Industries Association (EIA) is a trade association whose members consist of manufacturers of electronic equipment. Representatives from the member firms prepare documents for the purpose of standardizing and assuring compatibility of equipment. The material in this book adapted from Recommended Standards EIA-464, EIA-470, and EIA-478 has been used with the permission of the Electronics Industries Association, 2001 Eye Street, NW, Washington, DC 20006.

Catalog of EIA & JEDEC Standards & Engineering Publications ($5.00) plus the standards are available from the EIA Engineering Department at the above address. The EIA accepts phone orders at 202-457-4966. Note that standards previously designated "EIA RS-xxx" are now simply "EIA-xxx."

[4]*Private Branch Exchange (PBX) Switching Equipment for Voiceband Applications,* EIA Standard RS-464, Dec. 1979 (Washington, DC: Electronic Industries Association) and *Addendum No. 1 to EIA-464,* Aug. 1982. Includes PBX

performance objectives for loss, frequency response, intermodulation distortion, envelope delay difference, return loss, steady and impulse noise, longitudinal balance, crosstalk, and overload. A major revision of EIA-464 has been in the works for some time, but currently there is no release date. EIA-464: 149 pages, $36.00; Addendum: 172 pages, $26.00.

[5]*Telephone Instruments with Loop Signaling for Voiceband Application,* EIA Standard RS-470, Jan. 1981 (Washington, DC: Electronic Industries Association). Includes individual telephone set objectives for acoustical-to-electrical and electrical-to-acoustical loss and frequency response, longitudinal balance, and steady noise. RS-470 has been replaced by EIA-470A, July 1987, 90 pages, $35.00. The material from RS-470 used in this book remains unchanged in EIA-470A.

[6]*Multi-line Key Telephone Systems (KTS) for Voiceband Application,* EIA Standard RS-478, July 1981 (Washington, DC: Electronics Industries Association). Includes key-system objectives for acoustical-to-electrical and electrical-to-acoustical loss and frequency response, longitudinal balance, steady noise, and return loss. 95 pages, $34.00.

8-3 INSTITUTE OF ELECTRICAL AND ELECTRONICS ENGINEERS

The Institute of Electrical and Electronics Engineers (IEEE) is a professional and technical organization whose membership consists of individual engineers and scientists working in all areas of electrical engineering. Members working in committees use their expertise to write standards for use by other engineers. The *IEEE Standards Listing* is a no-charge catalog of standards published by the IEEE. The catalog and the standards are available from: IEEE Service Center, 445 Hoes Lane, Piscataway, NJ 08855. The IEEE accepts credit card orders at 201-981-0060.

[7]*IEEE Standard Test Procedure for Measuring Longitudinal Balance of Telephone Equipment Operating in the Voice Band,* ANSI/IEEE Standard 455-1985, 1985 (New York: The Institute of Electrical and Electronics Engineers, Inc.). Contains a measurement method (but no performance objectives) for longitudinal-to-metallic balance. IEEE 455 is often cited by other standards. 22 pages, $15.00.

[8]*IEEE Standard Methods and Equipment for Measuring the Transmission Characteristics of Analog Voice Frequency Circuits,* ANSI/IEEE Standard 743-1984, 1984 (New York: The Institute of Electrical and Electronics Engineers, Inc). Contains measurement methods only (no requirements or objectives) for level, frequency, noise, S/N ratio, envelope delay distortion, impulse noise, phase and gain hits, dropout, phase and amplitude jitter, return loss, peak-to-average ratio, intermodulation distortion, crosstalk, and slope. (IEEE 743-1984 replaces Bell's PUB 41009.) 60 pages, $16.00.

[9]*IEEE Standard Telephone Loop Performance Characteristics,* ANSI/IEEE Standard 820-1984, 1984 (New York: The Institute of Electrical and Electronics

Engineers, Inc). Includes definitions, measurement methods, and objectives for loss, frequency response, noise, and longitudinal balance of subscriber loops. 23 pages, $15.50.

8-4 BELL COMMUNICATIONS RESEARCH

Bell Communications Research, Inc. (Bellcore) is a relatively new organization. Prior to divestiture of the Bell System, AT&T and Bell Laboratories performed many services for the Bell operating companies. Postdivestiture, many of these same services have been performed by Bellcore—and by many of the same people (who transferred from Bell Labs). Bellcore is jointly owned by the seven Regional Bell Operating Companies.

One of the functions of Bellcore is to provide technical reference documents to the Bell companies and others. *Catalog of Technical Information* (no charge) plus the technical documents are available from: Bell Communications Research, Inc., Customer Services, 60 New England Avenue, Piscataway, NJ 08854. Credit card orders are accepted at the Bellcore Documentation Hotline (201-699-5800).

Some documents have not been updated since divestiture, so still contain references to the "Bell System" or "AT&T" in their titles. Such documents listed here are nevertheless available from Bellcore.

[10]*Data Communications Using Voiceband Private Line Channels,* Bell System Technical Reference PUB 41004, Oct. 1973 (New York: American Telephone and Telegraph Company). Includes theory and objectives on frequency response, envelope delay distortion, steady noise, loss, impulse noise, phase jitter, frequency shift, and echo. 68 pages, $20.75.

[11]*Transmission Parameters Affecting Voiceband Data Transmission—Description of Parameters,* Bell System Technical Reference PUB 41008, July 1974 (New York: American Telephone And Telegraph Company). A tutorial on subjects including frequency response, envelope delay distortion, noise, nonlinear distortion, incidental modulation, frequency shift, phase jitter, phase and gain hits, and dropout. 34 pages, $20.00.

[12]*Digital Channel Bank Requirements and Objectives,* Bell System Technical Reference PUB 43801, Nov. 1982 (Basking Ridge, NJ: American Telephone and Telegraph Company). PUB 43801 is the interface compatibility specification for D-type PCM carrier terminals. It replaces Bell's D3 compatibility specification of 1977. PUB 43801 includes voice-frequency requirements and test procedures for their verification. If you are involved with digital channel banks or related equipment at either the VF or digital interface, you should have access to this document. 184 pages, $27.75.

[13]*Notes on the BOC Intra-LATA Networks—1986* (TR-NPL-000275), copyright © 1986 Bell Communications Research, Inc. Material adapted from *Notes* is used with permission. Includes Intra-LATA network objectives for steady and impulse noise, loss, through and terminal balance, and overload. Describes nu-

merous other transmission parameters. Other sections of *Notes* cover switching, traffic, number plans, services, and maintenance. This document is the postdivestiture version of AT&T's *Notes on the Network* and the earlier *Notes on Distance Dialing*, both known as the "Blue Book." The current *Notes* is an excellent reference on everything in the network. 628 pages, $150.00.

8-5 RURAL ELECTRIFICATION ADMINISTRATION

The Rural Electrification Administration (REA) is part of the U.S. Department of Agriculture. The REA was formed many years ago to help bring electric power to rural areas. It also provided low-interest-rate loans to utilities as an inducement to serve sparsely populated rural areas with telephone service. In general, REA borrowers followed the REA's prescribed engineering practices and purchased equipment that meets REA standards. Over the years, REA's engineers have created quite a collection of documents that pertain to very basic and reliable telephone service.

The REA practices are listed in "Numerical Index," *REA Telecommunications Engineering and Construction Manual,* Section 102; REA specifications are listed in *REA Forms Catalog.* The index, the catalog, and other documents are available from Telecommunications Engineering and Standards Division, Rural Electrification Administration, U.S. Department of Agriculture, Washington, DC 20250. The REA's publications department accepts phone orders at 202-382-8674. There is no charge for REA documents.

[14]"General Specification for Digital, Stored Program Controlled Central Office Equipment," REA Form 522, May 1984 (Washington, DC: U.S. Department of Agriculture, Rural Electrification Administration). Includes digital end office objectives for impedance, loss, frequency response, overload, level tracking, return loss, longitudinal balance, steady and impulse noise, crosstalk, quantizing distortion, absolute delay, and envelope delay distortion. 95 pages.

[15]"General Specification for Common Control Central Office Equipment," REA Form 524, Jan. 1976 (Washington, DC: U.S. Department of Agriculture, Rural Electrification Administration). Includes analog switching office objectives for impedance, loss, return loss, longitudinal balance, steady and impulse noise, and crosstalk. 96 pages.

8-6 BOOKS

[16]Members of the Technical Staff, *Transmission Systems for Communications,* 5th ed. (Holmdel, NJ: Bell Telephone Laboratories, Inc., 1982). Includes theoretical treatments of loss, delay, hybrids, levels, steady noise, impulse noise, distortion, crosstalk, and echo. 921 pages.

[17]Bellamy, John, *Digital Telephony* (New York: John Wiley & Sons, Inc., 1982). Includes sections on pulse code modulation, quantizing noise, and idle channel noise. 526 pages.

[18]Brigham, E. Oran, *The Fast Fourier Transform* (Englewood Cliffs, NJ: Prentice-Hall, Inc., 1974). Contains a graphical presentation of the Fourier transform that can be followed without using heavy math. 252 pages.

[19] Brooks, John, *Telephone* (New York: Harper & Row, Publishers, 1975, 1976). An objective corporate history of AT&T and a social history of the telephone. 369 pages.

[20]Grossner, Nathan R., *Transformers for Electronic Circuits* (New York: McGraw-Hill Book Company, 1983). Contains a brief but good mathematical development of hybrid transformer operation. 467 pages.

8-7 TECHNICAL ARTICLES

[21]Batorsky, D. V., and M. E. Burke, "1980 Bell System Noise Survey of the Loop Plant," *AT&T Bell Lab. Tech. J.*, 63 (May–June 1984), 775-818. The 1980 Noise Survey evaluated 1256 subscriber loops for noise, longitudinal current, and longitudinal balance.

[22]Carey, M. B., H. T. Chen, A. Descloux, J. F. Ingle, and K. I. Park, "1982/83 End Office Connection Study: Analog Voice and Voiceband Data Transmission Performance Characterization of the Public Switched Network," *AT&T Bell Lab. Tech. J.*, 63 (Nov. 1984), 2059-2119. The 1982/83 End Office-Connection Study (EOCS) analyzed the transmission characteristics of 6141 end office-to-end office connections through the Bell System's predivestiture network. The connections were categorized by airline miles into short (0–180 mi), medium (181–720 mi), and long (721–2576 mi). This study, conducted just prior to divestiture, may serve as a reference against which postdivestiture networks can be compared. Transmission characteristics reported include loss, frequency response, steady noise, delay, S/N ratio, envelope delay distortion, peak-to-average ratio, intermodulation distortion, phase and amplitude jitter, frequency shift, impulse noise, phase and gain hits, and dropout.

[23]Manhire, L. M., "Physical and Transmission Characteristics of Customer Loop Plant," *Bell Sys. Tech. J.*, 57 (Jan. 1978), 35–59. A report of Bell's 1973 Loop Survey. Subscriber loops serving 1100 main stations were selected from throughout the Bell System and analyzed for loss, frequency response, return loss, and impedance.

[24]Miller, G., "The Effect of Longitudinal Imbalance on Crosstalk," *Bell Sys. Tech. J.*, 54 (Sept. 1975), 1227–1251. A theoretical development backed by laboratory and field measurements.

[25]Sartori, E., "Hybrid Transformers," *IEEE Trans. Parts Mater. Packag.*, PMP-4 (Sept. 1968), 59–66. A thorough mathematical development of hybrid transformer operation.

[26]Shennum, R. H., and J. R Gray, "Performance Limitations of a Practical PCM Terminal," *Bell Sys. Tech. J.,* 41 (Jan. 1962), 143–171.

8-8 PATENTS

Copies of patents are available for $1.50 each from the Patent and Trademark Office, Washington, DC 20231.

[27]"Intermodulation Distortion Analyzer," *United States Patent 3,862,380,* Jan. 21, 1975 (Washington, DC: U.S. Patent Office). This invention by Norris C. Hekimian and James F. Turner (assigned to Hekimian Laboratories, Inc.) covers the four-tone intermodulation distortion measurement now standard in the U.S. 14 pages.

8-9 SUMMARY

Table 8-1 is a summary list of the major documents referenced in this book. If you work with customer-provided or other terminal equipment, the documents marked in the first column would make a good starter library of standards. If, instead, you are involved with systems in the telephone network itself, the documents marked in the second column are more applicable. Items marked in the third column contain background theory useful to all areas of telephony.

TABLE 8-1 SUMMARY OF STANDARDS

Terminal equipment	Network	Theory	Short designation	Reference number	Subject
×			FCC 68	1	Terminal equipment
×			EIA 464	4	PBX
×			EIA 470	5	Telephone instruments
×			EIA 478	6	Key systems (KTS)
×	×	×	IEEE 455	7	Longitudinal balance
×	×	×	IEEE 743	8	Analog VF measurements
	×		IEEE 820	9	Subscriber loops
	×		PUB 41004	10	VF private channels
		×	PUB 41008	11	Analog VF parameters
	×		PUB 43801	12	Digital channel banks
	×		*Notes*	13	Intra-LATA networks
	×		REA 522	14	Digital end offices
	×		REA 524	15	Analog offices

INDEX